ISBN 978-3-662-37516-7   ISBN 978-3-662-38284-4
DOI 10.1007/978-3-662-38284-4
Softcover reprint of the hardcover 1st edition 1966

Bulletin

• INAUGURAL ISSUE •

Contents

Bulletin of Environmental Contamination and Toxicolog

AIMS AND SCOPE

The Bulletin of Environmental Contamination and Toxicology will provide rapid publication of significant advances and discoveries in the fields of pesticide residue research, air, soil, and water contamination and pollution, methodology, and other disciplines concerned with the introduction, presence, and effects of toxicants in the total environment.

Results of current research will be presented as brief reports providing information which is potentially useful to all individuals concerned with environmental contamination.

The articles will be free from restrictions imposed by purely scientific journals, particularly with respect to completeness of the studies reported and the attendant delays in publication.

Descriptions of new methods, procedures, or techniques shall be sufficiently detailed so as to permit direct application in other laboratories.

Review articles and obvious abstracts of papers forthcoming in other publications are not invited and probably will not be acceptable.

Articles suitable for inclusion shall be relatively short (less than 2,000 words) and will be prepared following specific instructions to permit reproduction by the photo-offset process from the original manuscript.

It is the hope of the Editorial Board that this Bulletin will provide a meeting ground for researchers who daily encounter problems related to the contamination of our environment and who welcome opportunities to share in new discoveries as they occur.

The Bulletin will be issued six times a year. This will be raised to 12 issues annually as demand increases.

Published bi-monthly by SPRINGER-VERLAG NEW YORK INC., 175 Fifth Avenue, New York, N. Y. 10010, Telephone (212) 673-9797. Six issues per year. Subscription price: $15 per year for institutions, $7.50 per year for individuals.

The Examination of Toxaphene
by Gas Chromatography

by Arthur Bevenue
Pacific Biomedical Research Center
University of Hawaii, Honolulu, Hawaii

and Herman Beckman
Agricultural Toxicology and Residue Research Laboratory
University of California, Davis, California

Considerable data have been published as tables on the retention times of various pesticides obtained by gas chromatography. Such data usually includes three or more "peaks" (or retention-time values) for the insecticide toxaphene. The paucity of published gas chromatographic curves on toxaphene, alone or in combination with other pesticides, can be explained by the continuum-like curves reproduced in Figure 1. The difficulty in classifying this type of curve qualitatively and estimating the amount quantitatively is obvious.

Toxaphene is a mixture of related compounds and isomers, with a chlorine content of 67 to 69 per cent. Therefore, well-defined separations of mixtures containing toxaphene cannot be successful because it is not a discrete compound (1). However, in an effort to find some reproducible, well-defined characteristic, or fingerprint, gas chromatographic studies were made of toxaphene alone, and in combination with technical grade DDT.

1

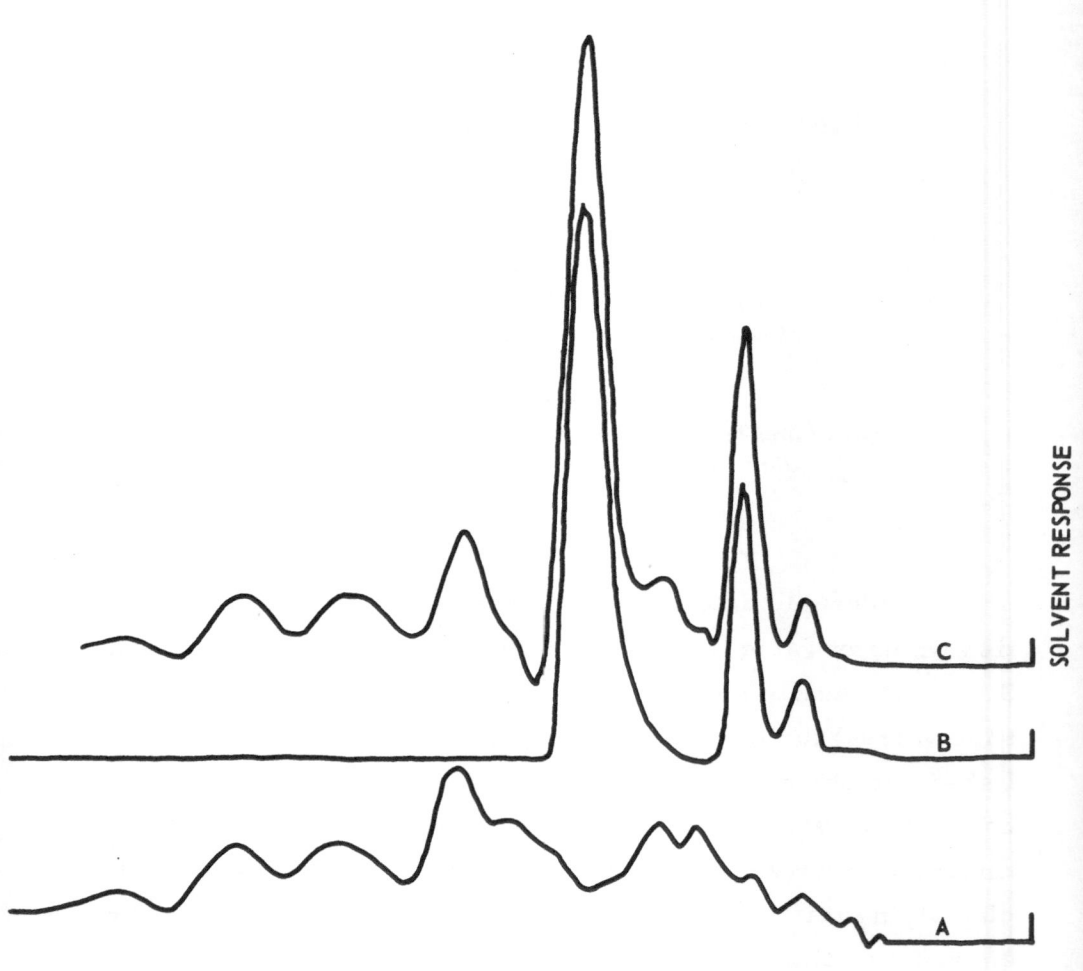

SOLVENT RESPONSE

ELECTRON CAPTURE DETECTOR RESPONSES TO: A – 7 NANOGRAMS
TOXAPHENE, B – 2.8 NANOGRAMS DDT, C – 7 NANOGRAMS
TOXAPHENE + 3.5 NANOGRAMS DDT

COLUMN: 4 x 1/4 WITH 5% QF - 1 ON CHROMASORB - W

Experimental

In an effort to find a characteristic fingerprint of toxaphene by the use of gas chromatography, the stationary liquid phase and the solid support of the gas chromatographic column were varied, both in the amount used and the type selected. The stationary phases included QF-1 (Silicone fluoro, FS 1265), Silicone Dow Corning 11, Silicone Dow Corning 200, and Silicone Gum Rubber, methyl, SE-30. In addition, columns were used containing mixtures of QF-1 and DC-200 oil in variable ratios. Column supports included Chromosorb G, Chromosorb W, and Chromosorb P, which had been variously treated with acid, hexamethyldisilazane, and/or dimethyldichlorosilane. Six different commercial types of gas chromatographs were used including the concentric and the parallel-plate type of electron capture detector (tritium radiation source), the pulse mode and the DC mode of operation, and also a coulometric type detector.

The detectable limit of toxaphene was dependent upon the newness (or cleanliness) of the detector and the efficiency of the column packing. Under ideal conditions, 2 nanograms of toxaphene could be observed with no difficulty. Using an instrument that had been exposed to considerable usage in pesticide residue analysis, the required sample size was of the order of 5 to 7 nanograms of toxaphene and 2 to 3 nanograms of DDT. Burke and Holswade (2) noted in their procedure when using a microcoulometric detector that it required about 9-fold the amount of toxaphene when compared to p, p'-DDT to attain one-half full scale recorder deflection.

Irrespective of the type of column or column packing used, no characteristic fingerprint for toxaphene could be obtained with one possible exception. One column material showed promise for a characteristic 3-peak fingerprint subsequent, timewise, to the p, p'-DDT peak. Using a 1/8 " x 5' glass column of 5% QF-1 on Chromosorb W, and assigning aldrin a reference

3

point of 1.00, o, p -DDT had a relative retention time of 2.7; p, p -DDT 4.1, and toxaphene had three "peaks" at retention time of 6.0, 7.0, and 8.0. Perhaps a different stationary liquid phase might be found that would improve upon and clearly define this portion of the toxaphene continuum.

The data indicates that the use of 1/8" columns containing Dow-11 or DC 200 oil at a column temperature not greater than 190° C. gave the best resolution for the DDT components. In the pesticide mixture, DDT is clearly defined in the presence of toxaphene if toxaphene does not exceed the amount of DDT present by an approximate factor of 3. Obviously, a gas chromatographic record of toxaphene is practically meaningless unless the background of the sample under examination is known; in many instances this is highly improbable, if not impossible to obtain. If the pesticide formulation originally used was a mixture, such as toxaphene-DDT, the picture is more complex. An experienced pesticide residue analyst may be competent in interpreting the data qualitatively, but to attempt to quantify such data may be more difficult and only an approximation. Proposals have been made to estimate the amount of toxaphene residue present in a sample by the so-called characteristic three-peak fingerprint comprising the latter part of the curve. This is demonstrated in Figure 1. However, until a more precise and sharper defined picture of this area is obtained by some improved means, this technique is highly empirical and subject to a large quantitative error.

Suggestions have also been made to rearrange the structure or to partially dehalogenate toxaphene, and DDT if present, by refluxing with a basic alcoholic solution. Gas chromatographic curves obtained from such solutions are then compared with curves obtained from untreated samples. Crosby and Archer showed that this procedure is not wholly successful, especially when toxaphene and DDT are in the same mixture. The dehydro-chlorinated toxaphene may show as many as 5 "peaks", by

4

electron capture gas chromatography, the primary peak being only about 2.0-2.5% as large as DDE, the dehydrochlorinated component of DDT (3).

In summary, the pesticide residue chemist has been placing increased reliance on gas chromatographic data for the identification of a pesticide residue. In the examination of a sample for toxaphene residue, such data is not reliable, either qualitatively or quantitatively. In particular, when State or other Regulatory agencies may wish to examine a shipment of produce suspected of excess toxaphene residue, the use of gas chromatography data alone for the basis for legal actions is an invitation for criticism and rebuttal. We believe the same thesis could be applied to the compounds chlordane and strobane. Until it can be shown by some new, and presently unknown, technique that toxaphene can be unequivocally identified, the gas chromatographic procedure for the determination of toxaphene, alone or in combination with other pesticides, is at best highly questionable. Further investigation into the "3-peak" phenomena at the latter part of the gas chromatographic curve may possibly produce a definitive fingerprint.

Acknowledgment

We acknowledge with thanks the cooperation of Mr. Fred Roth, Mr. Felix Erro, Mr. William Lewis, and Mr. R. A. Sweet , all of the California State Department of Agriculture, in supplying much of the data obtained from a variety of commercial types of gas chromatographic instrumentation.

References

1. H. F. BECKMAN and P. BERKENKOTTER, Anal. Chem. 35, 242 (1963)

2. J. BURKE and W. HOLSWADE, J. A. O. A. C. 47, 845 (1964)

3. D. G. CROSBY and T. E. ARCHER, Paper presented at the 148th National Meeting of the American Chemical Society, Chicago, September, 1964.

Drug Effects on Dieldrin Storage in Rat Tissue*

by J. C. STREET, M. WANG, and A. D. BLAU
Animal Science Department
Utah State University

The accumulation of certain insecticides or their
metabolites in the animal body following the consumption
of contaminated feed is one of the greatest detriments
to the use of insecticides in the production of food or
forage crops. Members of the chlorinated cyclodiene
insecticides, e.g., dieldrin, aldrin, endrin, and
heptachlor, represent the greatest potential problem
because of their relatively great storage rates in animal
tissues (1).

We recently obtained a marked reduction in dieldrin
storage in rat tissue in our laboratory by administering
DDT (2). This effect had apparent parallels in the well
established affects of DDT and certain other compounds
which stimulate the detoxification rates of a variety of
drugs (3,4). These similarities led us to test selected
drugs for their effectiveness in reducing dieldrin storage
in rats.

---

* Presented at the 150th meeting, American Chemical
Society, Atlantic City, September 15, 1965.

6

Experimental

The selected drugs, which are neither chemically nor pharmacologically related, are shown in figure 1[‡]. Aminopyrine and phenylbutazone are antipyretic and analgesic agents, heptabarbital is a sedative and tolbutamide is an antidiabetic. Each, however, stimulates the metabolic detoxification rate of some other drug. For example, aminopyrine (5), phenylbutazone (5), and tolbutamide (4) stimulate hexobarbital detoxification, phenylbutazone and aminopyrine (5) stimulate zoxazolamine detoxification and heptabarbital stimulates the metabolism of certain coumarin anticoagulants (6).

Individual female rats were fed a basal diet containing 1 ppm dieldrin (in one trial the feed contained 10 ppm dieldrin). The drugs were administered as feed additives or by i.p. injections. A negative control, consisting of a group treated only with dieldrin, and a positive control group which was treated with dieldrin plus DDT, were included in each experiment. Five rats were in each treatment group. The rats were sacrificed

---

[‡] The drugs used were generously supplied by the following companies: aminopyrine, Sterling-Winthrop Research Institute; heptabarbital and phenylbutazone, Geigy Research Laboratories; tolbutamide, The Upjohn Company.

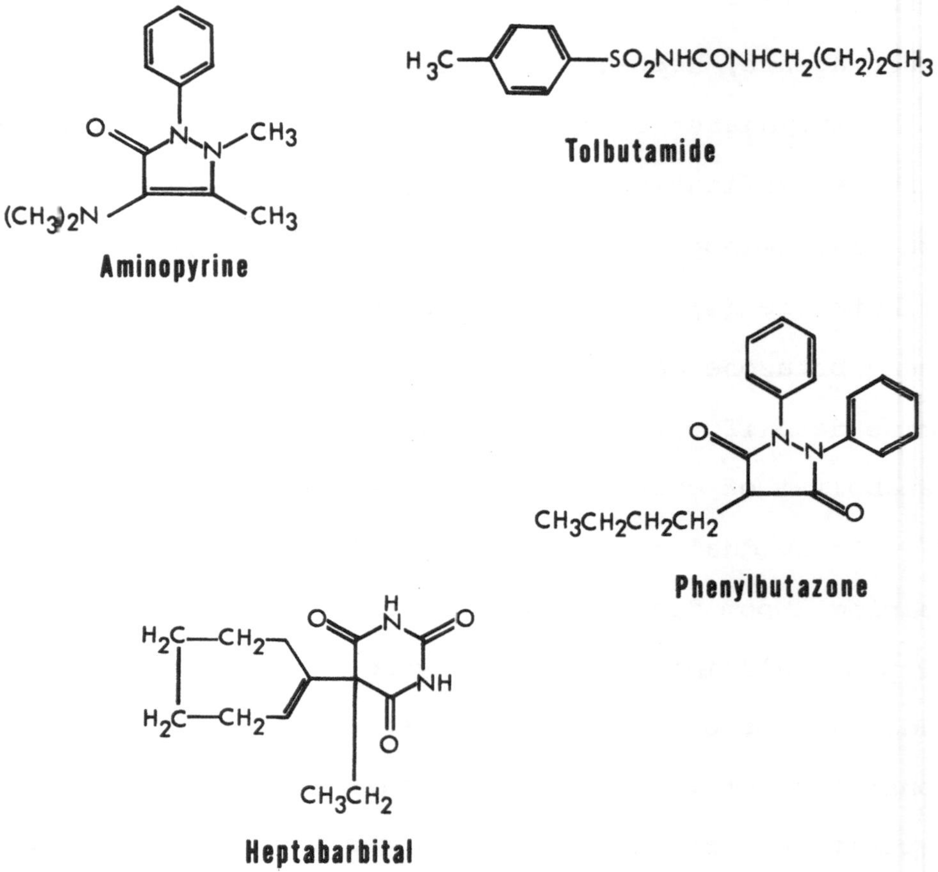

Figure 1. Drugs tested for effectiveness in reducing dieldrin storage in rat tissue.

after 10 days of treatment. Abdominal adipose tissue was analyzed for lipid content by a dichromate oxidation procedure. Residual dieldrin was determined by electron capture gas chromatography. These methods have been described in an earlier paper (2). Analysis of variance techniques were used to aid in interpreting results.

Results and Discussion

Results of the drug tests are summarized in tables 1-3. The concentration of dieldrin found in adipose tissue, as ppm in tissue lipids, is listed and also the storage reduction relative to control rats that received only dieldrin. Each drug caused a significant reduction in dieldrin storage when administered in the diet (tables 1 and 2). The groups treated with low drug levels had

TABLE 1

Drug effects on dieldrin storage. All rats received 1 ppm dieldrin in the diet continuously for 10 days. Dosages are listed as daily intakes per kilogram of body weight

Treatment	Tissue dieldrin		Storage reduction
	μg/g lipid	±SE	%
Control	7.53	0.89	
DDT, 4 mg/kg	2.06	0.34	72
Tolbutamide, 60 mg/kg	6.54	0.47	13
Tolbutamide, 290 mg/kg	3.20	0.72	57
Aminopyrine, 75 mg/kg	2.76	0.39	63
Aminopyrine, 350 mg/kg	1.40	0.09	81
Heptabarbital, 40 mg/kg	4.01	0.67	47
Heptabarbital, 225 mg/kg	1.50	0.07	80

TABLE 2

Effects of phenylbutazone on dieldrin storage. Series
A: All rats received 10 ppm dieldrin in the diet continuously for 10 days. Series B: All rats received 1
ppm dieldrin in the diet continuously for 10 days.
Dosages are listed as daily intakes per kilogram of
body weight

Treatment	Tissue dieldrin		Storage reduction
	μg/g lipid	±SE	%
A. Control	74.2	10.1	
Phenylbutazone, i.p.			
50 mg/kg, 4 days	75.7	1.5	0
DDT, i.p.			
20 mg/kg, 3 days	18.0	3.2	76
B. Control	9.38	0.67	
Phenylbutazone, fed			
90 mg/kg, 10 days	5.68	0.95	39
DDT, fed			
4 mg/kg, 10 days	1.97	0.33	79

dieldrin residues significantly lower than the control

group. The dieldrin residues were significantly higher,

however, than in the high drug dosage groups. This

suggests that normal dose-response relationships occur

with this phenomenon. Although the tests were not

designed for an accurate comparison of potencies, the

drugs appeared to rank in the following order: hepta-

barbital > aminopyrine ≫ tolbutamide. Phenylbutazone

was tested in a different trial and showed a potency

similar to tolbutamide. Heptabarbital, the most potent,

caused an 80 per cent reduction in tissue dieldrin concentration when given at the level of 225 mg/kg body weight for 10 daily doses. DDT, about 50 times more effective than heptabarbital, was far more potent than any of the drugs.

Some data indicated that the drug effect is either slowly activated or is of transient duration. In our initial experiment, phenylbutazone given i.p. for 4 days was ineffective but later proved to be active when fed for ten days (table 2). Possibly the longer time period was necessary in order for the effect to become fully activated. However, DDT was very effective after only 3 days of i.p. administration which is another contrast between the activity of DDT and the drugs. Somewhat similar observations were made with aminopyrine (table 3). When administered i.p. for 5 days it was less effective than when lower dosages were given the entire 10 days. Significant reductions of dieldrin storage were obtained, however, with the 5-day treatment schedule for all but the lowest aminopyrine dose level. Either the effect had diminished after the fifth day, or the mechanism was not fully activated during the 5-day period. A study of the relative potencies of these drugs and DDT and the duration of their effects is continuing in our laboratory.

TABLE 3

Effect of Aminopyrine on dieldrin storage. All rats received 1 ppm dieldrin in the diet continuously for 10 days. Dosages are listed as daily intakes per kilogram of body weight

Treatment	Tissue dieldrin		Storage reduction
	µg/g lipid	±SE	%
Control	6.36	0.50	
Aminopyrine, 10 days, i.p.			
25 mg/kg	6.30	0.62	0
50 mg/kg	4.92	0.64	22
75 mg/kg	3.66	0.15	42
Control	7.61	0.63	
Aminopyrine, 5 days, i.p.			
50 mg/kg	7.97	0.27	0
100 mg/kg	5.71	0.64	25
150 mg/kg	6.51	0.24	15

Others have recently reported similar results from the use of phenobarbital. Cueto and Hayes found reduced amounts of dieldrin-derived material in the fat of rats under chronic phenobarbital-dieldrin treatment as compared to rats treated with dieldrin alone (7). Koransky et al. reported a marked acceleration in elimination of BHC metabolites when rats were pretreated with phenobarbital (8).

These various drugs, including phenobarbital and DDT, are thought to stimulate detoxification reactions by inducing synthesis of hepatic microsomal enzymes which participate in the metabolism of lipid-soluble compounds (9). It is significant that dieldrin metabolism is affected by these non-specific agents and it is likely that the effects can be demonstrated with other members of the chlorinated cyclodiene insecticides and BHC, and in other animal species. These observations may lead to the development of agents which will safely reduce insecticide accumulation in the tissues of animals and man. Such agents might also be used for the treatment of individuals who may become over exposed to insecticides.

Summary

Selected drugs were tested for effectiveness in reducing dieldrin retention by rats. Female rats were fed diets treated to contain 1 ppm dieldrin. The drugs were administered as feed additives or by i.p. injections. The rats were sacrificed after 10 days of treatment and abdominal adipose tissue was analyzed for dieldrin using electron capture gas chromatography.

Heptabarbital (40 and 225 mg/kg rat/day), aminopyrine (75 and 350 mg/kg rat/day), tolbutamide (60 and 290 mg/kg rat/day), and phenylbutazone (90 mg/kg rat/day) were effective as feed additives in reducing tissue dieldrin. Heptabarbital was the most effective and reduced the concentration of tissue dieldrin by 80 per cent at the higher dose level. In comparison, DDT

13

(4 mg/kg rat/day) effected a 72 per cent reduction.
A contrast with DDT was also observed in trials with
i.p. administration of drugs and DDT. In those trials,
the duration of the DDT action was apparently greater
than that of the drugs.

We suggest that suitable drugs might be used to
reduce insecticide accumulation in the tissues of animals
and man, and for treatment of individuals after over
exposure to insecticides.

Acknowledgements

This work was aided by USDA regional research funds
(project W-45) and by USPHS research grant EF-00543
from the Division of Environmental Engineering and Food
Protection. Gifts of drugs by the Sterling-Winthrop
Research Institute, Geigy Research Laboratories, and
The Upjohn Company are deeply appreciated.

LITERATURE CITED

1. CLABORN, H. V., R. D. RADELEFF, and R. C. BUSHLAND, U.S.D.A. Bulletin, ARS-33-63, (1960)
2. STREET, J. C., Science 140, 1580 (1964)
3. HART, L. G. and J. R. FOUTS, Proc. Soc. Exp. Biol. Med. 114, 388 (1963)
4. REMMER, H., Proc. First International Pharmacological Meeting, Vol 6, p 235, (1962), Pergamon Press, Oxford.
5. CONNEY, A. H., C. DAVISON, R. GASTEL and J. J. BURNS, J. Pharmacol. Exp. Therap. 130, 1 (1960)
6. DAYTON, P. G., Y. TARCAN, T. CHENKIN, and M. WEINER, J. Clin. Invest., 40, 1797 (1961)
7. CUETO, C., JR., and W. J. HAYES, JR., Toxicol. and Appl. Pharmacol., 7, 481 (1965)
8. KORANSKY, W., J. PORTIG, H. W. VOILAND and I. KLEMPAU, Naunyn-Schmiedebergs Arch. Exp. Path. Pharmak., 247, 49 (1964)
9. BRODIE, B. B. and R. P. MAICKEL, Proc. First International Pharmacological Meeting, Vol. 6, p 299, (1962), Pergamon Press, Oxford.

A Rapid Analytical Method for Persistent Pesticides in Proteinaceous Samples

by D. G. CROSBY and T. E. ARCHER
Agricultural Toxicology and Residue Research Laboratory
University of California, Davis, California

A recent investigation (1,2) of the relationship between
dietary intake of pesticides and subsequent tissue levels required development of a very rapid, simple, and highly sensitive method for the detection and estimation of chlorinated hydrocarbon insecticides in proteinaceous materials. The following
procedure has been applied successfully to about 3000 samples of
this type in our laboratory and has been particularly useful for
analysis of DDT and its relatives in small samples of bovine
milk and blood.

Experimental

Chemicals and Equipment. All chemicals were reagent grade.
Pesticides were analytical standards supplied by the manufacturer
or purified in our own laboratory, and the reagent grade solvents
were redistilled shortly before use. The gas chromatograph was
an Aerograph Model 600B (Wilkens Instrument Co.) equipped with an
electron capture detector and a Leeds and Northrop Model G 1 mv.
recorder. The most satisfactory chromatographic column was a
9' x 1/8" stainless steel tube packed with 60/80 mesh HMDS-
treated Chromosorb W containing 5 per cent Dow 710 silicone oil
and 5 per cent SE-30 gum rubber. The 12" section of the column
nearest to the injection port was packed with 20/30 mesh calcium
carbide for removal of traces of water and ethanol. Nitrogen
carrier gas (40 p.s.i. 60-100 ml./min.), a column temperature of
240-250°C, and an instrument attenuation of 1X gave best results.

Presented at the 148th National Meeting, American Chemical
Society, September, 1964.

Bulletin of Experimental Contamination & Toxicology,
Vol. 1, No. 1, 1966, published by Springer-Verlag New York Inc.

A recorder chart speed of 60"/hr. was convenient, and the peak areas were measured with a Polar planimeter.

The dehydrohalogenation reagent was prepared fresh daily by dissolving (for each sample) 5 g. of C.P. potassium hydroxide in 3 ml. of distilled water and adding, with stirring, 17 ml. of ethanol.

Analysis of Milk. The milk was warmed to 40°C., mixed well, and a subsample (generally 10 ml.) was pipetted into a glass-stoppered Pyrex test tube (about 50 ml. capacity) containing 20 ml. of the KOH reagent. The tube was stoppered, shaken, heated in a water bath at 75-80°C. for 15 min., cooled, and 10 ml. of pentane was added to its contents. After shaking for 4 min., the layers were allowed to separate, and an aliquot of the upper (hydrocarbon) layer was withdrawn with a syringe and injected directly into the chromatograph. If the pentane and alkali phases did not separate immediately, the addition of 10 ml. of distilled water hastened the separation. In this instance, one µl. of the pentane solution was equivalent to 1 mg. of milk. The amount of pesticide in the aliquot was determined by comparing the area under the peak on the strip-chart to those obtained with appropriate dehydrohalogenated standard solutions (containing 0.5 ng. in the case of DDT).

Analysis of Blood. Citrated, oxalated, or heparinized blood (0.1 - 1.0 ml.), thoroughly mixed, was introduced into 2 ml. of the KOH reagent and then analyzed according to the procedure for milk with the exception that only 1 ml. of pentane was added for extraction.

Analysis of Fat and Flesh. The sample (1-2 g.) was weighed into the 5 ml. cup of a VirTis micro homogenizer and blended with 5 ml. of pentane. The homogenate was transferred to a 500 ml. boiling flask with about 200 ml. of an equivolume mixture of pentane and ether, boiled under reflux for 1 hr., cooled, filtered, and the solvent removed in vacuo on a rotating evaporator. The residual extractives were weighed, 10 ml. of pentane added, and

17

an aliquot equivalent to 0.5 g. of extractives was transferred to a glass-stoppered test tube. The solvent was removed in a stream of air, and 20 ml. of KOH reagent was added. Analysis then was conducted according to the milk procedure.

Alternatively in each of these analyses, a round-bottomed flask could be substituted for the test tube, and the alkaline dehydrohalogenated mixture could be transferred to a separatory funnel for pentane extraction. This was useful when relatively large samples were to be analyzed.

Results and Discussion

As shown in Table 1, a number of chlorinated hydrocarbon insecticides may be estimated in proteinaceous samples by this "dehydrochlorination" procedure. Toxaphene, heptachlor, and endrin provided well-resolved, multiple peaks, one of which predominated. Although a few almost coincident peaks occurred, those of an intensity greater than 5 per cent of the DDT peak were distinctly separated.

TABLE 1.
Retention Times of Dehydrochlorinated Pesticides

Compound	Retention Time in Minutes (Peak Intensity[a])		Compound	Retention Time in Minutes (Peak Intensity[a])	
Lindane	0.4	(13.9)	DDD (p,p')	6.7	(64.0)
Toxaphene I	2.7	(0.4)	Endrin II[b]	7.2	(1.1)
Toxaphene II	3.2	(0.4)	Toxaphene V	8.1	(1.0)
Heptachlor I[b]	4.1	(7.3)	DDT (p,p')	8.3	(100.0)
Aldrin	4.6	(97.3)	DDE (p,p')	8.3	(100.0)
Toxaphene III[b]	4.8	(2.4)	Endrin III	9.7	(0.5)
Heptachlor II	5.1	(3.1)	Dieldrin	10.3	(5.8)
Endrin I	5.2	(0.5)	Heptachlor epoxide [c]		0.0
Toxaphene IV	6.7	(0.9)	Thiodan [c]		0.0

[a] DDE 100, [b] Principal peak, [c] no peak at 30 ng.

The major part of our work involved measurement of total DDT and relatives (p,p'-DDT, o,p'-DDT, p,p'-DDE, and DDD) in bovine milk, bovine and human blood, poultry eggs, and the fat/flesh from cattle, deer, geese, pheasants, and fish. Extracts of plant products including alfalfa hay, feed formulations, and pasture grass generally required the addition of a Florisil cleanup step to minimize pigments. DDT could not be distinguished from DDE by this method, of course, and the retention times of o,p'-DDE and the dehydrohalogenation product of p,p'-DDD were coincident: analysis for the total DDT group therefore was greatly simplified. The very high column temperature gave convenient retention times, sharp peaks, and excellent column life without detector fouling. A single column functions well for more than a year, and detector sensitivity was not affected when the column was well conditioned.

Analytical sensitivity and accuracy were very satisfactory. Standard recoveries were essentially quantitative, and, in the ranges most frequently encountered, results were reliable to within ± 3 per cent. (Table 2). Other experiments showed that the

TABLE 2

Probable Error (P) in Analyses of Several Sample Types

Sample	No. of Samples	DDT[a] (p.p.m. ±P)	± P (Per cent)
Whole milk	5	0.645 ± 0.006	0.9
Colostrum	5	0.0031 ± 0.00003	1.0
Bovine Blood	5	0.408 ± 0.0009	2.2
Bovine Blood	5	0.00302 ± 0.0004	1.2
Bovine Blood	5	0.00061 ± 0.00002	2.9

[a] Total of DDT and related compounds

error was almost equally divided between the sampling-extraction-cleanup steps and the chromatography-measurement steps. During our investigation, Schafer, et al (3) described a very similar procedure in which peak height was used for quantitation; we found, however, that measurement of peak areas at these very low levels offered a notable increase in accuracy over the other method.

The principal advantages of this method lie in (a) a degree of simplicity which permits a large volume of high-protein and/or high-fat samples to be handled routinely, (b) the very effective cleanup afforded by the hot alkali, and (c) the chemical conversion of a variety of "persistent" pesticides into derivatives more readily separated and detected than the parent compounds. We are indebted to R. C. Laben and S. A. Peoples for milk, blood, and tissue samples; to H. F. Beckman for helpful discussions; to Nels Larsen and Eugene Whitehead for technical assistance; and to Margaret Schafer for a prepublication copy of her excellent manuscript.

References

1. R. C. LABEN, S. A. PEOPLES, and D. G. CROSBY, Proc. Third Ann. Conf. on the Use of Agric. Chemicals in Calif., Davis, Calif., Jan. 14, 1964.

2. R. C. LABEN, T. E. ARCHER, D. G. CROSBY and S. A. PEOPLES, J. Dairy Sci. 48, 701 (1965).

3. MARGARET SCHAFER, K. A. BUSCH, and J. E. CAMPBELL, J. Dairy Sci. 46, 1025 (1963).

Secretion of DDT in Milk
by Fresh Cows[1,2]

by W. H. Brown,[3] J. M. Witt,[4] F. M. Whiting,[5] and J. W. Stull[3]

Environmental contamination of animal feed by pesticides
applied onto nearby crops is considered to be the major source
of pesticide residues found in milk. However, examination of
the pesticide input from these sources frequently does not account for the level of pesticide residue found in the corresponding milk. Attempts to explain the discrepancy between the
amount of pesticide residue found in milk and the amount which
one can predict should be found based on many feeding studies
(1, 2, 3, 4) has led many persons to suggest that there may be a
relationship between these unexplainably high pesticide residues
in milk and the stage of lactation in the cow, i.e. that cows
which have just come fresh secrete more pesticide in their milk
fat than during later stages of lactation. The rationale which

[1] Arizona Agricultural Experiment Station Technical Paper No. 1078

[2] This work was supported in part by Grant No. EF-00627-01 from
the U.S.P.H. and a Grant from the United Dairymen of Arizona
and The Arizona Federated Milk Producers.

[3] Department of Dairy Science, University of Arizona, Tucson
[4] Department of Entomology, University of Arizona, Tucson
[5] United Dairymen of Arizona, Tempe, Arizona

21

prompted this suggestion is based on the fact that a cow loses considerable body weight, presumably stored body fat, during the first weeks following parturition. Since cows store insecticide in their body fat at a level near that secreted in the fat of milk (1, 2, 5, 6) it appeared reasonable that depletion of body fat would release the previously stored insecticide and make an unusually large amount of insecticide available for secretion in the milk. This theory has been sporadically tested by analysis of single samples of milk from fresh cows or groups of fresh cows and compared to samples from cows in later stages of lactation in the same herd which have received the same dietary exposure to insecticides. This type of testing has given conflicting results as to whether fresh cows excrete more insecticide in milk fat than those in later stages of lactation (Ariz. State Dept. Health., pers. comm.; Calif. State Dept. Agric. pers. comm.; unpublished data, this laboratory). Laben et al. (6) give data which show a decline in the level of DDT in the milk fat immediately after parturition for a period of 40 weeks in a sequence of samples taken from the same cows. However, all but the control group of cows were also declining from an exposure of DDT which had been terminated 30 days prior to parturition and this decline would mask any effect due to changes in fat mobilization. Their data presented for the control group of cows, which received only that DDT which unavoidably appeared in the hay ration (ca. 0.05 ppm), shows that there was little or no pattern of decline for 20 weeks following parturition, but that there was a decline from ca. 1.5 ppm to ca. 0.25 ppm from the 20th to the 30th post-partum weeks

22

and this decline was followed by a rise to 0.4 ppm by the 40th
week.

The supposition that fresh cows may contribute a dispropor-
tionate share of insecticide to the pooled herd milk has been
primarily concerned with the possibility of a gross disproportion
occurring in the first 2 or 3 post-partum weeks rather than for
the first half of the milking period. It was the purpose of this
study to examine the insecticide level in the milk in the very
early stage of lactation by sampling with sufficient frequency to
detect short term fluctuations that would be masked by weekly
composites. This milk was sampled at every milking for 14 days
and sampled six times per week and composited into two samples
per week following this earliest lactation period.

Eight Holstein cows from the University of Arizona dairy
herd were used as experimental animals. Six of the animals were
in their 2nd lactation, one in her 3rd, and one in her 5th lactation. The cows were maintained on the same source of feed during
the last stages of their lactation, the dry period, and the post-
partum testing period, although the amounts of hay and grain were
adjusted according to Morrison's standards. The cows received
hay only during the dry period which extended for 43 to 87 days,
with an average of 61 days. The animals did not receive any special dose of DDT prior to or during the experimental period. The
only known insecticide intake was from the DDT which constituted
an unavoidable contamination of the feed. The hay contained an
average of 0.01 ppm of DDT and its degradation products and the
grain contained 0.05 ppm. Milk samples were taken twice daily

for the first 14 days following parturition and analyzed individually. The daily average is presented in table 1.

TABLE 1

Secretion of Pesticides and Milkfat in the Milk of Fresh Cows

Time after Part.	DDT	DDE	DDD	Total	Fat	Milk	Fat
(days)	PPM in Milkfat				(%)	(kg/day)	(kg/day)
0	0.13	0.49	0.01	0.73	3.4	10.1	0.34
1	0.12	0.56	0.01	0.69	4.2	15.6	0.66
2	0.14	0.54	0.02	0.70	3.3	20.4	0.67
3	0.08	0.62	0.01	0.71	4.3	18.2	0.78
4	0.09	0.55	0.02	0.66	4.5	20.2	0.91
5	0.09	0.63	0.01	0.73	5.0	19.8	0.99
6	0.11	0.50	0.00	0.61	4.6	19.8	0.91
7	0.10	0.54	0.01	0.65	5.5	21.8	1.20
8	0.11	0.51	0.02	0.64	5.6	18.0	1.01
9	0.12	0.45	0.01	0.58	4.8	22.0	1.06
10	0.11	0.57	0.02	0.70	4.0	21.6	0.86
11	0.10	0.47	0.01	0.58	4.0	23.8	0.95
12	0.11	0.56	0.01	0.68	4.1	23.2	0.95
13	0.13	0.54	0.02	0.69	3.5	26.4	0.92
14	0.09	0.54	0.02	0.65	3.4	23.8	0.81
15-17	0.11	0.58	0.01	0.70	3.3	25.8	0.85
18-21	0.11	0.62	0.02	0.75	3.5	26.4	0.92
22-24	0.12	0.56	0.01	0.69	3.4	27.4	0.93
25-28	0.10	0.56	0.01	0.67	3.3	25.0	0.83
29-31	0.10	0.61	0.01	0.72	3.3	25.0	0.83
32-35	0.13	0.64	0.02	0.79	3.1	26.2	0.81
36-38	0.13	0.61	0.02	0.76	3.1	27.4	0.85
39-42	0.10	0.50	0.01	0.61	3.3	25.0	0.83
43-45	0.10	0.58	0.01	0.69	3.3	25.2	0.83
46-49	0.12	0.58	0.02	0.72	3.0	26.4	0.79
50-52	0.14	0.53	0.02	0.69	3.0	25.0	0.75
53-56	0.10	0.49	0.01	0.60	3.0	25.2	0.76
57-59	0.11	0.51	0.01	0.63	3.0	25.1	0.75

From 14 to 59 days the milk was sampled once a day and composited for analysis semi-weekly. The analysis was carried out by electron capture gas chromatography and DDT, DDE, and DDD were measured separately (7).

24

The data are presented in table 1 as the average response from the eight cows studied. It can be seen by simple inspection that, although the level of DDT and its degradation products in the milk fat range from 0.58 to 0.79 ppm, there is no consistent trend either up or down of the level of DDT and its products over the 59 day period studied. The daily production of milk fat followed a slight upward trend until the 7th post-partum day and then decreased slightly but not significantly. The mean values and standard deviations for the 59 day experimental period are presented in table 2.

TABLE 2

Average Secretion of Pesticides and Milk Fat in the Milk of Fresh Cows

	Number of Observations	Average	Standard Deviation
DDT (ppm in milk fat)	165	0.119	0.145
DDE (ppm in milk fat)	165	0.549	0.123
DDD (ppm in milk fat)	164	0.018	0.022
Total pesticide (ppm in milk fat)	162	0.680	0.174
Percent fat in milk	210	3.65	1.306
Daily milk production (kg)	210	23.66	0.641
Daily fat production (kg)	210	0.86	0.129

The data were analyzed by methods described by Snedecor (8) to determine the standard deviation and the significance of any trends.

The standard deviations express the variation encountered between cows and the analytical error. The coefficients of variance of the chemical analyses were DDE, 26.2%; DDD, 93.5%; and DDT, 46.0%. The level of detection of DDT (0.1 ppm in the fat;

0.040 ppm in whole milk) and DDD (0.02 ppm in the fat; 0.001 ppm
in whole milk) was so close to the level of the reagent blank when
expressed in equivalent terms (0.03 to 0.05 ppm of milk fat), that
a high standard deviation and coefficient of variance are to be
expected. The variations of pesticide level and fat production
found between cows and within cows with regard to time showed no
consistent trend and were not of sufficient magnitude to be significant. Although many persons have rationalized that the fresh
cow could uniquely contribute to the level of pesticide in the
herd milk not only because of the release of pesticide stored in
the body fat, but also because of a much higher level of production of milk fat during the early stage of lactation, this suggested higher level of fat production was not substantiated by
this data nor has this concept been substantiated by any previous
work (9) despite the continued belief that this phenomenon occurs.
There is an approximate 30 percent decrease in the amount of fat
produced at the end of the entire lactation period as compared to
the early stages, but this is a gradual decline over the entire
period.

Although the absence of trends was noted for the averaged
response of the eight cows, there was an increase in the volume
of milk production for the first ten days of lactation for three
of the eight cows, and there was considerable variation (but no
consistent trends) in percent milk fat for all of the individual
cows for the first three or four days, but not in the weight of

milk fat produced or in the level of pesticide residue in the
milk fat. It was also shown that the cows in this study lost
an average of only 37 lbs. body weight in the first month after
calving. In the second month there actually was an average gain
of 12 lbs. per cow.

It is interesting to note that although the feed for the
cows on this test had a contamination of DDT (and its degradation
products) which averaged out at 0.023 ppm, the milk fat had a
residue of 0.68 ppm which is 30 times the residue level which
would be predicted from the level in the diet (1, 2, 3, 4). Although the ratio of the level of pesticide in the dose to the
level of pesticide in the milk was near to unity in the work of
Laben et al. (6), their control group of animals also showed the
same 30:1 disparity detected here (0.05 ppm in feed; 1.5 to 0.35
ppm in milk fat). This disparity poses a difficult problem in
the management of herds producing milk carrying a level of pesticide residue near 2.0 ppm in the milk fat but which have been consuming feed carrying a pesticide residue level near or below 0.1
ppm. It is shown by the data presented here that this problem is
not caused by a unique release of stored pesticides by cows just
coming fresh.

Acknowledgments

The assistance of Dr. Henry Tucker, Associate Director of
the Numerical Analysis Laboratory, the University of Arizona, is
greatly appreciated. The assistance of the late Mrs. N. Lord and

the Mrs. M. Milbrath, and G. Shaw in the processing of the samples is also greatly appreciated.

References

1. Ely, R. E., L. A. Moore, R. H. Carter, H. D. Mann, and R. W. Poos, J. Dairy Sci. 35, 266. (1952)

2. Shepherd, J. B., L. A. Moore, R. H. Carter, and F. W. Poos. J. Dairy Sci. 32, 549. (1949)

3. Williams, S., P. A. Mills, and R. E. McDowell. J.A.O.A.C. 47, 1124. (1964)

4. Zweig, G., L. M. Smith, S. A. Peoples, and R. Cox. J. Agr. Food Chem. 9, 481. (1961)

5. Bruce, W. N., R. P. Link, and G. C. Decker. J. Agr. Food Chem. 13, 63. (1965)

6. Laben, R. C., T. E. Archer, D. G. Crosby, and S. A. Peoples. J. Dairy Sci. 48, 701 (1965)

7. Witt, J. M., F. M. Whiting, W. H. Brown, and J. W. Stull. Manuscript submitted to J. Dairy Sci. (1966)

8. Snedecor, G. W. Statistical Methods. (1956) Iowa State College Press, Ames, Iowa.

9. Jenness, Robert and Stuart Patton. Principles of Dairy Chemistry. (1959). John Wiley & Sons, Inc., New York.

Confusion of Identification of o,p′-Kelthane as Heptachlor in Orange Rind Extractives

by W. E. WESTLAKE, R. T. MURPHY, and F. A. GUNTHER
Department of Entomology,
University of California Citrus Research Center
and Agricultural Experiment Station, Riverside, California

Gas and thin-layer chromatograms of Valencia and Navel orange rind stripping solutions from mature fruits treated commercially on the trees early in the season with the acaricide 1,1-bis (p-chlorophenyl)-2,2,2-trichloroethanol (Kelthane) had revealed the presence of an unknown compound not present in control samples. It had a tlc R_f value and a glc retention time practically identical with those of heptachlor and was reported as apparent heptachlor in a "market" survey by an independent laboratory using published methods, yet today heptachlor is not used in citriculture. The data herein presented prove conclusively that this compound was not heptachlor and show that it has retention and other characteristics of o,p′-Kelthane, a compound known to be present in small amounts in technical grade Kelthane. Microcoulometric glc analysis of the stripping solution further showed the isolate contained organo-chlorine. Soil samples from under the treated trees contained both isomers of Kelthane but again no detectable heptachlor.

Method

Principles

Samples of Valencia and Navel orange rinds were equilibrated with mixed hexanes. Soil samples were equilibrated with 2:1 hexane-isopropyl alcohol, the alcohol was washed out, and the hexane stripping solutions were dried over anhydrous sodium sulfate. A residue was partitioned into acetonitrile and then back into hexane by addition of water to the acetonitrile phase. After drying, the hexane was concentrated and chromatographed through a Florisil column. Aliquots of the eluate fractions were analyzed by electron-capture gas chromatography. All 3 types of column packings used to separate the organochlorine residues gave different retention times relative to aldrin, allowing partial characterization by reference to purified standards. Portions of the eluates were also fractionated by glc and by tlc for further characterization by R_f values, spectral absorption, and polarography.

Determinations

Extraction and Partition -- Five-hundred grams of finely chopped orange peel were equilibrated with 1000 ml. of mixed hexanes by end-over-end tumbling for 1 hour. After filtration,

200 ml. of the stripping solution was extracted with two 50-ml. portions of acetonitrile. Combined acetonitrile extracts, diluted with 300 ml. of water, were extracted with three 50-ml. portions of hexane. Combined hexane extracts were dried over anhydrous sodium sulfate and concentrated to 10 ml. through Snyder columns. Soil samples of 1000 g. each were equilibrated with 2:1 hexane-isopropyl alcohol by tumbling for 1 hour. The alcohol was washed out with distilled water, the hexane solutions were dried over anhydrous sodium sulfate, then concentrated to about 100 ml. through Snyder columns.

Column Chromatography -- Ten grams of Florisil was added to a Shell-type glass column (2) and washed with 100 ml. of hexane. A sample was washed onto the Florisil layer with two 10-ml. portions of hexane followed by an additional 50 ml. The eluate was collected as a single fraction beginning with the addition of the sample; heptachlor added to equivalent portions of control orange rind stripping solutions was eluted in this fraction.

Then 100 ml. of 6% diethyl ether in hexane was passed through the column and collected. Both o,p'- and p,p'-Kelthane were eluted in this fraction. Further elution with 25% ether in hexane did not remove more of these compounds from the Florisil.

Gas chromatography -- Gas chromatographic analyses of aliquots of these column fractions were with an electron-capture detector (Aerograph Pestilyzer Model 680). Most evaluations involved a 2-foot borosilicate, 1/8-inch O.D. column packed with 100/200 mesh acid-washed HMDS, Chromosorb-W support coated with 5% SF-96, at 150° C. Verifications were on a 29-cm. Teflon, 0.034-inch I.D. column packed with 60/80 mesh Teflon coated with 1.0% Apiezon L, also at 150° C. Nitrogen at 20 ml./min. was maintained through each column. Further aliquots were put through an Aerograph Autoprep 705 glc unit with electron-capture detector. A 2-foot x 1/8-inch stainless steel column packed with 3.4% QF-1 and 6.2% DC-200 on Gas Chrom Q was used, at 180° C.; nitrogen flow was 75 ml./min.

Thin-layer chromatography -- Aliquots were concentrated to near dryness and spotted on fluorescent silica gel thin-layer plates. After development with 5:1 ether-benzene, R_f values were determined. Appropriate areas were scraped off, the compounds were extracted, and the resulting solutions were gas chromatographed.

Spectrophotometry -- Infrared spectra were recorded with a Perkin-Elmer 21 instrument equipped with sodium chloride optics and 0.3-ml. cavity cells of 5-mm. light path. Ultraviolet spectra were recorded on a Beckman DK-2 instrument and silica cells of 1-cm. light path.

32

Polarography -- Polarograms were determined at 25° C. oscillographically with a Davis Southern Analytical Differential Cathode Ray Polarotrace, Type A.1660A equipped with an amalgamated silver wire reference electrode and Gajan micropolarographic cells. Reference solutions contained 5 to 10 μg. of compound/ml. of equal volumes of 95% ethanol and 0.2 $\underline{M}$ tetramethylammonium bromide solution.

Results and Discussion

Breakdown of purified $\underline{p},\underline{p}'$-Kelthane to $\underline{p},\underline{p}'$-dichlorobenzophenon in 15 to 85% yields in gas chromatographic systems has been reported by Gunther et al. (1). Both $\underline{p},\underline{p}'$- and $\underline{o},\underline{p}'$-Kelthane were converted in good yields to their corresponding dichlorobenzophenones in the 3 gas chromatographic systems used in this study.

Technical grade Kelthane could not be gas chromatographed reproducibly until it had been passed through the Florisil column cleanup procedure. The fraction eluted by 6% ether in hexane contained compounds with the same glc retention times as those from purified $\underline{o},\underline{p}'$- and $\underline{p},\underline{p}'$-Kelthane.

The similarity of the retention times of heptachlor and $\underline{o},\underline{p}'$-Kelthane was shown by the formation of a distinct doublet peak when a mixture of the two compounds was chromatographed on the SF-96 or Apiezon L columns. The mixed QF-1-DC-200 column

33

completely resolved the two peaks despite their close relative retention times (0.77 and 0.94, respectively, when aldrin = 1.00).

The 6% ether fraction of control orange rind samples, processed as described, produced no peaks in the area of interest. This same fraction from subject field-treated oranges, however, showed the glc presence of two compounds with retention times identical to those of o,p'- and p,p'-Kelthane. When o,p'- and p,p'-Kelthanes were added to this fraction each corresponding peak height was increased with no indication of new doublet formation. When heptachlor was added, however, a new doublet was formed with the SF-96 and Apiezon L columns and distinct heptachlo and o,p'-Kelthane peaks resulted when the QF-1-DC-200 column was used. When both heptachlor and o,p'-dichlorobenzophenone were added to the subject orange rind extractives, using the combinatio column a distinct heptachlor peak and an o,p'-Kelthane (o,p'-dichlorobenzophenone) peak of correspondingly increased height resulted.

As stated earlier, heptachlor would have been eluted from the Florisil column in the hexane fraction. Gas chromatography as above of this fraction from the subject orange rind samples did not yield a peak with a retention time even remotely similar to that of ehptachlor.

Table I gives the retention times, relative to aldrin, for the 3 gas chromatographic columns used. The most notable feature

34

is the efficiency of the QF-1-DC-200 column in separating heptachlor

and o,p'-Kelthane (o,p'-dichlorobenzophenone) despite the very small

difference in retention times. This column, however, would not

separate aldrin and o,p'-dichlorobenzophenone, and an analyst might

easily be led to believe that aldrin was present in a routine

examination of a Kelthane-treated sample. This again emphasizes

the necessity for confirming gas chromatographic analyses by a

more specific method.

TABLE I

Effect of column packing on retention times relative to aldrin

Compound	Time relative to aldrin		
	SF-96	Apiezon L	QF-1 + DC-200
Heptachlor	0.83	0.73	0.77
o,p'-Kelthane	0.89	1.10	0.94
o,p'-Dichlorobenzophenone	0.89	1.10	0.94
Aldrin	1.00	1.00	1.00
p,p'-Kelthane	1.20	1.80	1.16
p,p'-Dichlorobenzophenone	1.20	1.80	1.16

Attempts to confirm the identity of the unknown compound in

the subject oranges by infrared spectrophotometry failed because

less than 10 μg. of pooled material was available, and both glc and tlc fractionation let some interfering substances through. Comparative ultraviolet spectrophotometry of these same fractions indicated the probable presence of o,p'-dichlorobenzophenone. Similarly, micropolarography was inconclusive because of overlap of reduction potentials (half-wave potentials were -0.66 v and -0.85 to -0.89 v for the two peaks of p,p'-Kelthane, -1.25 to -1.2 for p,p'-dichlorobenzophenone, and -1.29 to -1.31 v for the o,p'-ketone).

The remainder of the pooled glc fractions, analyzed in a Dohrmann microcoulometric gas chromatograph, contained chlorine in the amount estimated from the previous analyses using the electron-capture detector.

Soil samples from the Kelthane-treated plots, analyzed by glc in the same manner as the rind, contained compounds that produced peaks identical with those for o,p'- and p,p'-Kelthane but no peak corresponding to heptachlor. Microcoulometric glc analyses again showed that organochlorides were present. Thin-layer chromatography of the soil extracts failed to separate the unknown from p,p'-Kelthane with the solvent system used. Thus, chromatograms developed on silica gel plates with 5:1 ether-benzene showed an elongated area with 3 distinct zones which were carefully removed, eluted, and gas chromatographed. Zone 1, with an R_f range of 0.66 to 0.73, contained the major part of a compound having a retention time identical with that of o,p'-Kelthane

Zone 2, with an R_f range of 0.54 to 0.66, contained a small amount of the unknown plus p,p'-Kelthane. Zone 3, with an R_f range of 0.47 to 0.54, gave no peaks in the area of interest. The R_f range for purified p,p'-Kelthane was 0.64 to 0.71 which agrees with the above observed distribution of this compound between zones 1 and 2. The R_f range for o,p'-Kelthane was 0.65 to 0.72. A 1:1 mixture of the 2 isomers afforded R_f 0.62 to 0.72. The R_f range for heptachlor in this system was the same as that of this mixture, making it impossible to distinguish between it and technical grade Kelthane (a mixture containing up to several % of the o,p'-isomer).

Conclusions

Both the glc and the tlc data prove that the unknown compound in the rind of Kelthane-treated oranges was not heptachlor. While the characterization is not absolutely conclusive, there can be no reasonable doubt that it is o,p'-Kelthane, known to be present in technical grade Kelthane.

Other workers, using glc, reported p,p'- and o,p'-Kelthane in the ratio of 100:1 in 21 market fruit samples from different areas of Southern California. By analogy with DDT, if technical grade Kelthane is assumed to contain up to 20% of the o,p'-isomer this ratio found on fruits treated months previously means that on and in oranges the o,p'-isomer has much the longer half-life

(RL_{50}) of the two isomers. On the other hand, approximately equal half-lives are indicated for the two isomers in soil.

This brief study emphasizes the need for checking on gas chromatographic data by a more specific method and clearly demonstrates again that improper interpretation of data may be avoided by using at least two columns possessing different retention characteristics for the compounds of interest.

Acknowledgments

The authors are indebted to Rohm and Haas Company, Philadelphia, Pa. for an analytical sample of o,p'-Kelthane, to D. White and F. Hearth for assistance in the laboratory, and to L. K. Gaston for valuable suggestions. Paper no. 0000 University of California Citrus Research Center and Agricultural Experiment Station, Riverside.

References

(1) F. A. GUNTHER, J. H. BARKLEY, R. C. BLINN, and D. E. OTT, Pesticide Research Bull., Stanford Research Inst. 2; 3 (1962).

(2) SHELL CHEMICAL CORP., "Determination of Endrin in Animal Tissues, Eggs, Butter, and Milk by Total Chloride Method", pp. 1-32, Aug. 13, 1956.

Bulletin

Contents

Bulletin of Environmental Contamination and Toxicology

AIMS AND SCOPE

The Bulletin of Environmental Contamination and Toxicology will provide rapid publication of significant advances and discoveries in the fields of pesticide residue research, air, soil, and water contamination and pollution, methodology, and other disciplines concerned with the introduction, presence, and effects of toxicants in the total environment.

Results of current research will be presented as brief reports providing information which is potentially useful to all individuals concerned with environmental contamination.

The articles will be free from restrictions imposed by purely scientific journals, particularly with respect to completeness of the studies reported and the attendant delays in publication.

Descriptions of new methods, procedures, or techniques shall be sufficiently detailed so as to permit direct application in other laboratories.

Review articles and obvious abstracts of papers forthcoming in other publications are not invited and probably will not be acceptable:

Articles suitable for inclusion shall be relatively short (less than 2,000 words) and will be prepared following specific instructions to permit reproduction by the photo-offset process from the original manuscript.

It is the hope of the Editorial Board that this Bulletin will provide a meeting ground for researchers who daily encounter problems related to the contamination of our environment and who welcome opportunities to share in new discoveries as they occur.

The Bulletin will be issued six times a year. This will be raised to 12 issues annually as demand increases.

Published bi-monthly by SPRINGER-VERLAG NEW YORK INC., *175 Fifth Avenue, New York, N. Y. 10010, Telephone (212) 673-9797.*

Conversion of a Dohrmann Microcoulometric Gas Chromatograph to a Convenient and Rapid "Total Chloride" Unit

by Francis A. Gunther and James H. Barkley
Department of Entomology
University of California Citrus Research Center
and Agricultural Experiment Station, Riverside, California

In quantitative residue investigations of the persisting organochlorine pesticides in foodstuffs there is frequent need to establish magnitudes of the total organically bound chlorine present. "Total" residue values of this sort can be interpreted in terms of the most toxic compound thought to be present, or with pesticides that are mixtures initially (such as BHC, Tedion, toxaphene, or chlordane) or that degrade to mixtures (such as DDT, DDE, DDD, and DDA) they can represent that mixture existing as a persisting residue. When organic chloride detection methods are coupled with gas chromatography, as in the Dohrmann Microcoulometric Gas Chromatograph, a high degree of specificity in residue methodology is achievable, as discussed elsewhere (1, 2). This instrumentation also provides the presently very necessary detection in the microgram region.

Some pesticides such as toxaphene present broad smears by

Bulletin of Experimental Contamination & Toxicology,
Vol. 1, No. 2, 1966, published by Springer-Verlag New York Inc.

ordinary glc, but these smears can be compressed into usable peaks by using very short columns (3, 4) or -- in some instances and with the Dohrmann instrument -- by by-passing the use of a column entirely and using the injection block to release volatile components, the combustion unit to convert covalently bonded chlorine to chloride ion, and microcoulometry to measure the total chloride present regardless of the time required. Thus, a "short-circuited" Dohrmann instrument becomes a very rapid and convenient total chloride apparatus. A final advantage to this type of operation is that if conversion from normal column to no-column operation could be achieved quickly, reversibly, and conveniently without cool-down, the instrument uniquely becomes capable of providing precise evaluations of column efficiency: operation without a column would establish efficiency of combined operations of injection, volatilization, combustion, and microcoulometric titration, then quick conversion to column operation without changing any of the other parameters would afford column efficiency data by simple subtraction. Most gas chromatographable pesticides and pesticide alteration products containing either chlorine or sulfur (2, 5) (e.g., Ethion, Tedion, Trithion) could be evaluated this way.

This communication describes how to convert a Dohrmann glc module into a unit for either normal or column by-pass use. The change from pre-chromatographed combustion to direct combustion is made by screwdriver operation of a by-pass valve located in the column oven of the Dohrmann instrument.

A drawing of the injector block, by-pass valve, and column is shown in Fig. 1.

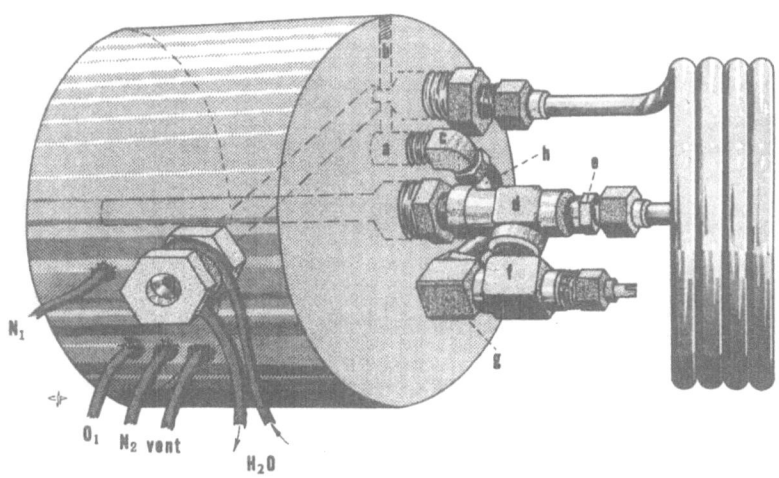

Fig. 1. Modified Dohrmann injection block-column assembly: a = new tapped hole with access to injection port through new hole b (later welded closed at surface of block), c = fitting connected to new by-pass valve f through new by-pass h, d = new T-fitting, e = enlarged connector, and g = new elbow. All fittings are stainless steel. Valve f has Teflon packing.

CONSTRUCTION

Materials (see Fig. 1): c = stainless steel elbow, one arm 1/4" O.D. Swagelok, one arm 1/4" NPT (Swagelok #400-2-4-316); d = stainless steel tee, 1/4" NPT (Cajon #4-T-316); e = stainless steel male connector, one end 1/4" O.D. Swagelok, one end 1/4" NPT (Swagelok #400-1-4-316); f = stainless steel regulating valve with Teflon packing, angle pattern with 1/4" NPT inlet and outlet connections (Whitey #NM-4-A-316); g = stainless steel female connector, one arm 1/4" O.D. Swagelok, one arm 1/4" NPT (Swagelok #400-8-4-316); and h = 1/4" O.D. stainless steel tubing.

Assembly: Remove the column entrance and exit connectors from the block of a Dohrmann Microcoulometric Gas Chromatograph. Drill 7/16" hole a 3/4" deep in the face of the block halfway between the connector ports, then tap it for 1/4" NPT. Drill 3/16" hole b as shown in Fig. 1, about 3/4" from the face of the block so that it interconnects hole a and the bottom reduced portion of the column entrance port; then weld hole b closed at the surface of the block. Screw standard Swagelok elbow c tightly into hole a using Teflon Thread Sealing Compound. The two previously removed connectors are lubricated with this sealing compound and replaced as tightly as possible.

To reduce as much as possible height of fittings above the face of the block grind off side and one end "shoulder" of tee d and the shoulder on the pipe thread end of elbow g; extend the threads as far as possible into these cut-down fittings.

Temporarily assemble pieces d, f, and g; cut the 1/4" O.D. tubing h, and bend and fit it to connect elbows c and g with Swagelok fittings. Remove handle of valve f and slot the shaft for screwdriver operation. Using the sealing compound on all threads assemble the pieces d, then f to g, then f to d, and then h to c and g. Drill connector e to 1/4" I.D. then insert it into d using the sealing compound.

The final assembly should be sufficiently compact to allow

installation of a 12-foot column without interfering with the
oven fan. Test all joints with soap solution after the oven has
reached 250° C.

OPERATIONAL CHARACTERISTICS

Install a 2-foot, 1/4" O.D. column packed with acid-washed
Chromosorb P containing 5% SF-96. Operating parameters are:
column nitrogen 170 cc./min., sleeve nitrogen 160 cc./min.,
primary oxygen 625 cc./min., oxygen injected through the hydrogen
port and into a second platinum gauze (to achieve more efficient
combustion and to eliminate the necessity for venting solvent)
120 cc./min., column temperature 180° C., block temperature 225°
C., and furnace temperature 800° C. A good test sample consists
of a mixture of 0.2 or 0.4 µg./µl. each of purified lindane, aldrin, and dieldrin in 95% ethyl alcohol; the Cl^- contributions
of the latter are then lindane 0.292, aldrin 0.233, and dieldrin
0.223 µg./µl. for a total Cl^- value of 0.748 µg./µl. Under the
above conditions and with column operation clean peaks are obtained at 2.7, 5.3, and 9.5 min., respectively. From 10 to 50
µl. of clean solvent provides no background under these conditions, even with no venting.

In Table I are reproduced typical results of column versus
no-column operation. To illustrate relative throughput times,
the recommended mixture of lindane, aldrin, and dieldrin requires
about 10 min. for total elution without column by-pass, whereas
only 0.4 min. is required with column by-pass. This unit has been
in intermittent operation for six years with no leakage past the

valve stem. Average recoveries from nearly a hundred routine samples of various organochlorine pesticides and many columns averaged about 90% for the by-pass open and about 75% for the by-pass closed. Samples containing less than about a microgram of organically bound chlorine are erratically high in apparent recoveries with either mode of operation.

TABLE I

Typical relative efficiencies of column (by-pass closed) versus no-column (by-pass open) operation with a standard mixture of lindane, aldrin, and dieldrin (see text for details).

Added		Cl^- Found,	Av. % recovery	
Vol., µl.	Chloride,[a]/µg.	µg.	By-pass open	By-pass closed
Standard 0.04 µg. each/µl.				
2	0.2	0.2±0.1	120	----
2	0.2	0.2±0.1	---	100+
6	0.5	0.6±0.2	112	----
6	0.5	0.6±0.2	---	120
10	0.8	0.8±0.1	107	----
10	0.8	0.6±0.2	---	77
20	1.6	1.4±0.1	90	----
20	1.6	1.1±0.2	---	70
Standard 0.4 µg. each/µl.				
5	3.9	3.3±0.1	85	----
5	3.7	2.7±0.2	---	73
10	7.5	6.8±0.2	90	----
10	7.5	4.6±0.4	---	71

TABLE I - Continued

Added		Cl⁻ Found,	Av. % recovery	
Vol., µl.	Chloride,[a] µg.	µg.	By-pass open	By-pass closed
Standard 0.4 µg. each/µl.				
20	15.0	12.8±0.3	85	----
20	15.0	10.5±0.5	---	70
40[b]	15.0	12.9±0.4	86	----
40[b]	15.0	11.3±0.4	---	75

[a] As organically bound chlorine in the standard mixture of various concentrations in 95% ethyl alcohol.

[b] Standard 0.2 µg. each/µl.

REFERENCES

(1) D. M. COULSON, Adv. Pest Control Research 5, 153 (1962).

(2) C. C. CASSIL, Residue Reviews 1, 37 (1962).

(3) J. M. WITT, G. F. BAGATELLA, and J. C. PERCIOUS, Stanford Research Institute Pesticide Research Bull. 2(1), 4 (1962).

(4) L. C. TERRIERE, U. KIIGEMAGI, A. R. GERLACH, and R. L. BOROVICKA, J. Agr. Food Chem. 14, 66 (1966).

(5) J. A. CHALLACOMBE and J. A. McNULTY, Residue Reviews 5, 57 (1964).

Paper No. 1658, University of California Citrus Research Center and Agricultural Experiment Station, Riverside.

Residual Nature of Certain Organophosphorus Insecticides in Grain Sorghum and Coastal Bermudagrass[1]

by H. W. Dorough, N. M. Randolph and G. H. Wimbish

Department of Entomology
Texas A&M University, College Station, Texas

Residue studies were conducted on several insecticides being evaluated for the control of the two-spotted mite, Tetranychus telarius (Linn.), on grain sorghum and the fall armyworm, Spodoptera frugiperda (Smith), on Coastal bermudagrass. On grain sorghum, the residual persistance of dimethoate, ethion, azinphosmethyl and methyl parathion was determined while similar studies were performed with Imidan, N-(mercaptomethyl) phthalimide-S-(O,O-dimethylphosphorodithioate, and trichlorfon on Coastal bermudagrass.

Treatment

Grain Sorghum. The first applications were made when the sorghum grain was still in the dough stage. Emulsifiable formulations of the insecticides were applied at the rate of 0.5 lbs. of toxicant per four gallons of water per acre. A high clearance spray rig with No. 3 Teejet nozzles spaced at 20-inch intervals along a horizontal boom was used to apply the insecticides under a pressure of 60 psi. For this treatment, the nozzles were

---

[1] Contribution No. TA-5350, Texas Agricultural Experiment Station, accepted for publication.

46

about six inches above the sorghum heads with one nozzle directly over the plants and another in the middle of each row.

A different nozzle arrangement was used for the second applications made on the same plots seven days after the first treatment. To obtain more thorough plant coverage, 0.5 lbs. of insecticide per nine gallons of water per acre were applied as described above except that the nozzle positioned between the rows was replaced by two nozzles at the end of a 20-inch drop extending between each row. One of the nozzles was directed to the side of the plants in one row and the other nozzle adjusted to spray the sides of plants in the adjoining row. Each plant, then, received spray from the top and directly from both sides.

Samples for residue analysis were collected at 0, 3, and 6 days after the first treatment and 0, 3, 6, and 14 days following the second application. Zero-day samples were taken within one hour after treatment. The heads were kept separate from the forage samples and the grain removed for analysis.

Coastal Bermudagrass. Trichlorfon and Imidan sprays were prepared from emulsifiable concentrate formulations and applied with a high clearance sprayer at a rate of 1.0 lb./a. and 0.75 lbs./a., respectively. Conditions under which they were applied were the same as described for the first treatment of the grain sorghum. Samples were collected at 0, 1, 3, 7 and 14 days following treatment by cutting the stems just above ground level.

Analysis

Grain Sorghum. All of the residue analyses were based on described colorimetric methods with some modifications in the extraction and clean-up techniques. The detection of dimethoate depended on alkaline hydrolysis to yield thioglycolic acid which was determined colorimetrically using the original Folin's uric acid method as described by Giang and Schechter, (1). Ethion was determined spectrophotometrically as a copper salt complex of its hydrolysis product, diethyl phosphorodithioic acid (2). Techniques for clean-up and detection of azinphosmethyl were taken from Adams (3) and Anderson (4) which involved alkaline hydrolysis to anthranilic acid, diazotization and coupling with N-(1-naphthyl) ethylenediamine to produce color. Methyl parathion residues were detected using the colorimetric method of Averell and Norris (5).

Coastal Bermudagrass. Trichlorfon residues were determined with a Barber-Colman Series 5000 Gas Chromatograph equipped with a sodium thermionic detector (6). Chromatography conditions were as follows: A six foot, four mm i.d. glass column was packed with 30% XF-1150 on 60/80 mesh acid-washed Chromosorb W. A 10 mg. plug of glass wool was loosely packed into the injector port end of the column to achieve maximum thermal breakdown of trichlorfon to its phosphate moiety (7). Nitrogen was used as the carrier gas at a pressure of 20 psi. Air and hydrogen was supplied to the burner at a pressure of 44 and

48

20 psi, respectively. All supply cylinders were fitted with a length of capillary tubing and the expressed pressures represent the reading of a pressure regulator at the supply cylinder. Applied potential was obtained from a 300 volt battery and the baseline current was held constant at $5X10^{-9}$ with an electrometer setting of 10^{-8} ampere full scale. Temperatures were as follows: column 180°C, injector 280°C and detector oven 215°C. The recorder chart speed was 0.25 inches per minute. Under these conditions, trichlorfon had a retention time of two minutes.

Extraction of trichlorfon from the bermudagrass samples was accomplished using a procedure similar to the one described by Anderson (7). Frozen samples were run through a Hobart Food Chopper and 100 g. subsamples processed for analysis. Each subsample was ground in a Waring blender with 300 ml. of 0.1 N H_2SO_4 for five minutes and the homogenate filtered through cheesecloth. Following centrifugation of the filtrate for 10 minutes at 2000 rpm, the supernatant was decanted into a one liter separatory funnel. Sixty-eight g. of sodium chloride was added, dissolved by shaking and then the aqueous mixture extracted twice with 200 ml. portions of ether. The combined ether extracts were evaporated to about 10 ml., 50 ml. of benzene added and the water in the sample removed with anhydrous sodium sulfate. The dried benzene-ether mixture was reduced to a total volume of one ml. Two microliters of this solution was injected into the gas chromatograph.

Imidan residues were determined using colorimetric and gas chromatographic methods. One-hundred g. of chopped bermudagrass were homogenized in 400 ml. of benzene, filtered through glass wool into a one liter separatory funnel and the filtrate extracted with 100 ml. of water. The benzene phase was separated from the water layer, dried with anhydrous sodium sulfate, filtered and a volume equivalent to five g. of plant material removed for further clean-up. This involved passing the benzene extract through a column containing five g. of a charcoal adsorbent mixture (1:1 Darco G-60 and Hyflo Super-Cel) and eluting the Imidan with 70 ml. of a 1:1 benzene-chloroform mixture. The effluent was then evaporated to a volume of about 0.5 ml. and the residue transferred to a 60 ml. separatory funnel with two 10 ml. portions of hexane and five ml. of acetonitrile. After vigorous shaking, the acetonitrile layer was removed and the hexane re-extracted with 5 ml. of fresh acetonitrile. The acetonitrile extracts were combined.

Color was developed in the exact manner as described in the colorimetry section of the report by Batchelder and Patchett (8). Briefly, this method was based upon the conversion of Imidan to anthranilic acid which was then coupled with 3-methyl-2-benzothiazolone hydrazone to produce a magenta colored product.

For analysis using gas chromatography the 10 ml. of acetonitrile was reduced in volume to one ml. and two microliters injected into the instrument. Chromatographic conditions different than those described for trichlorfon analysis are as follows:

50

the column was packed with 1.5% SE 30 on 80/100 mesh acid-washed

Chromosorb P and the nitrogen pressure was reduced to 16 psi.

Air and hydrogen pressure were 44 and 25 psi, respectively.

Operating temperatures were as follows: column 200°C, injector

250°C and detector oven 220°C. The retention time for Imidan

was 2.5 minutes.

Results and Discussion

Grain Sorghum. After slight modifications in some of the

techniques, recovery of each insecticide from both forage and

grain samples exceeded 80 percent. Practical sensitivity levels

of the methods, based on the amount of interferring materials

from untreated samples, were as follows: dimethoate, 0.20 ppm;

ethion, 0.15 ppm; and methyl parathion, 0.20 ppm.

Although all four organophosphates were applied at the rate

of 0.5 lbs./a., the initial residue deposits varied with each

compound. This variation was more evident in the forage samples

of the first treatment where the dimethoate and methyl parathion

treated samples had residues of approximately 4.5 ppm immediately

after treatment while the ethion and azinphosmethyl samples con-

tained residue of 2 to 3 ppm (Table 1). Residues on the forage

just after the second application varied from about 8 ppm for

dimethoate to 14 ppm for azinphosmethyl. These variations in

the actual amounts of active material deposited on the plant did

not seriously affect the residue study since the rate of dissi-

pation of the toxicant present was the major factor under con-

sideration.

51

TABLE 1

P.P.M. Insecticide Residues on Sorghum Forage at Various Intervals After Treatment 1/

Days After Treatment	P.P.M.			
	Dimethoate	Ethion	Azinphos-methyl	Methyl Parathion
First Treatment				
0	4.65	2.13	2.90	4.80
3	2.21	0.79	0.76	0.36
6	0.56	0.55	0.64	ND2/
Second Treatment				
0	8.13	11.45	13.60	12.56
3	2.96	4.60	4.41	1.61
6	1.07	1.32	1.71	0.49
14	0.64	0.36	0.99	ND

1/ All materials applied at a rate of 0.5 lbs./a.
    First application June 26, 1964; Second application
    July 3, 1964.
2/ None detected

Dissipation rates for dimethoate, ethion and azinphosmethyl from the forage were very similar. About 60 percent of the material deposited had disappeared by three days after either treatment. By six days, approximately 20 percent of the applied dose was still present and extending the persistance study to 14 days after the second treatment revealed that less than 10 percent of the insecticides were still on the plant portion of the treated sorghum. The rate of dissipation did not appear to be maintained after the residues declined below the 1.0 ppm level. These small quantities of residues slowly disappeared and resulted in over 0.50 ppm in the plant six days after treatment.

Analysis of forage samples from methyl parathion-treated grain sorghum showed that the residual life of this insecticide was very short. Of those residues found immediately after the first or second treatment, 4.80 ppm and 12.56 ppm respectively, less than 10 percent was present after three days. No residues were detected six days after the first application or 14 days after the second.

More consistent residues were deposited on the grain (Table 2). Deposits were similar for each insecticide within a treatment and fairly consistent after both the first and second applications. In general the residues were on the order of 2 to 4 ppm. A major exception was grain from the second application of methyl parathion where only 0.80 ppm residues were detected on these 0-day samples and none at all at the later sampling dates. There was no apparent explanation for these low residues since the corresponding plant samples contained 12.56 ppm methyl parathion residues.

Dissipation of the four organophosphates was slightly slower from the sorghum grain than from the plant. Again, however, dimethoate, ethion and azinphosmethyl reacted in a similar manner. Residues of approximately 2 to 4 ppm of these insecticides deminished to almost half these levels in three days.

Unlike the forage samples, the residues on the grain did not appear to persist after levels of less than 1.0 ppm were present. This was shown by the fact that only ethion residues could be detected on the grain after 14 days.

TABLE 2

P.P.M. Insecticide Residues on Sorghum Grain at Various Intervals After Treatment. 1/

| Days After Treatment | P.P.M. | | | |
	Dimethoate	Ethion	Azinphos- methyl	Methyl Parathion
First Treatment				
0	2.50	3.93	3.13	3.14
3	1.60	2.02	1.45	0.36
6	0.68	0.99	0.95	ND2/
Second Treatment				
0	1.78	2.96	2.44	0.80
3	0.95	1.54	0.96	ND
6	0.66	1.09	0.57	ND
14	ND	0.85	ND	ND

1/ All materials applied at a rate of 0.5 lbs./a.
First application June 26, 1964; Second application July 3, 1964.
2/ None detected

Although the second application methyl parathion-treated grain samples contained unexplainably low residues, the results of the residual nature of this material on grain after the first treatment clearly showed it to be of short residual life. Only the three-day samples had detectable amounts of residues present and these were less than 0.40 ppm.

Residues on sorghum forage after the first and second applications clearly demonstrated that the same amount of active material per acre resulted in different residue levels when applied in a different manner. Insecticide deposits were as much as four times greater in concentration after the second treatment.

The fact that the residues on the grain after both treatments were almost the same suggests that adjustment of nozzles could be important in preventing insecticide residues on certain portions of a crop. In a crop such as sorghum it is likely that insects attacking only the stalks and leaves could be controlled with minimum insecticide contamination of the sorghum grain.

Coastal Bermudagrass. Attempts were first made to determine trichlorfon using electron capture gas chromatography (7). However, the procedure could not be used in the present study because interfering materials with the same retention time as trichlorfon were present in untreated Coastal bermudagrass.

Since the electron capture method was dependent upon the quantitative thermal breakdown of trichlorfon to chloral, the compound actually detected when trichlorfon was injected, it appeared likely that the phosphate moiety could be determined using a sodium thermionic detector. This was actually the case as trichlorfon was easily detected upon injection of as little as

10 nanograms. At this concentration a peak height of six mm. was observed and the response was linear up to 300 nanograms. Injections of untreated bermudagrass extracts did not show any contamination peaks in the trichlorfon area. Unlike the electron capture detector, this method could not determine chloral or trichlorethanol residues.

Very good recoveries were obtained when trichlorfon was added to 100 g. of untreated bermudagrass at four different levels ranging from 0.1 to 3.0 ppm. The maximum recovery was 120 percent, a minimum of 88 percent and an average of all recoveries of 97 percent.

Table 3 shows the residues present on Coastal bermudagrass at intervals up to 14 days following trichlorfon treatment. Almost half of the material deposited at treatment, 59.12 ppm, had diminished after only one day. After this, the rate of dissipation declined and residues of slightly more than five ppm remained in the plants 14 days after treatment.

Imidan was determined using colorimetry and sodium thermionic methods of detection. By colorimetric analysis, recoveries were greater than 88 percent and exceeding 97 percent when the residues were detected by sodium thermionic gas chromatography. Several extraction steps necessary in the former method but eliminated in the latter probably accounted for the increased recovery. Certainly, the comparative recoveries indicated that the sodium thermionic method of detecting Imidan residues could be used with confidence.

TABLE 3

Residues on Coastal Bermudagrass following Trichlorfon and Imidan Treatment 1/

Days After Treatment	P.P.M.	
	Trichlorfon	Imidan
0	59.12	37.93 (26.89)[2]
1	29.37	32.66 (24.91)
3	16.02	23.89 (19.81)
7	8.04	15.79 (14.15)
14	5.74	8.31 (11.05)

1/ Trichlorfon and Imidan applied at 1.0 and 0.75 lbs./a., respectively on September 20, 1964.
2/ Numbers in parentheses are ppm as determined colorimetrically.

Sodium thermionic and colorimetric determination of Imidan residues on Coastal bermudagrass reflected the same findings observed in the recovery experiments. In all but the 14 day samples, the gas chromatography method resulted in higher residues, indicating better recoveries (Table 3). Both procedures showed residues of a similar magnitude and served as a check for each other in demonstrating the residual persistence of Imidan in Coastal bermudagrass.

Imidan deposits on the plants at the time of treatment, approximately 38 ppm, dissipated slower than trichlorfon residues. With Imidan, however, the rate of dissipation continued fairly constant throughout the 14 day test period, with a half-life of between five and six days.

References

1. P. A. GIANG and M. S. SCHECHTER, J. Agr. Food Chem. $\underline{11}$, 63 (1963).
2. J. R. GRAHAM and E. F. ORWALL, J. Agr. Food Chem. $\underline{11}$, 67 (1963).
3. J. M. ADAMS, Chemagro Corporation, Report No. 13534, September, 1961.
4. C. A. ANDERSON, Chemagro Corporation, Report No. 6499, January, 1961.
5. P. R. AVERELL and M. V. NORRIS, Anal. Chem. $\underline{20}$, 753 (1948).
6. L. GIUFFRIDA, JAOAC $\underline{47}$, 293 (1964).
7. R. J. ANDERSON, Chemagro Corporation, Report No. 14770, November, 1964.
8. G. H. BATCHELDER and G. G. PATCHETT, Stauffer Chemical Co., Report No. RR-65-82, June, 1965.

Metabolites of Methyl- and Dimethylcarbamate Insecticide Chemicals as Formed by Rat Liver Microsomes

by E. S. Oonnithan and J. E. Casida
Division of Entomology and Acarology
University of California, Berkeley, California

Critical interpretation of toxicology and residue studies on an insecticide chemical is possible only when the metabolism of the compound is understood. Biologically active metabolites may be formed, particularly if the modification on the molecule occurs at a site other than the toxophoric grouping. In this respect, metabolites of carbamates formed by mechanisms other than initial hydrolysis at the carbamic ester site may be of importance. Mammals, and microsomal enzymes of liver, carry out many types of hydroxylation reactions including, among others, aromatic hydroxylation, aliphatic hydroxylation, $\underline{N}$-dealkylation, $\underline{O}$-dealkylation, and sulfoxidation (1, 2). Groupings which are potentially susceptible to such hydroxylation reactions are

Bulletin of Experimental Contamination & Toxicology,
Vol. 1, No. 2, 1966, published by Springer-Verlag New York Inc.

present in methylcarbamate insecticide chemicals.

There is some information in regard to the identity of metabolites formed in mammals or by microsomal enzyme systems from carbaryl (1-naphthyl methylcarbamate), Baygon (2-isopropoxyphenyl methylcarbamate), Pyramat (6-methyl-2-propyl-4-pyrimidinyl dimethylcarbamate), and certain dimethylcarbamates which yield N-methyl, N-hydroxymethylcarbamates (3-8). The in vivo fate of the radiocarbon is known for 10 variously C^{14}-labeled methyl- and dimethylcarbamate insecticide chemicals, but, other than $C^{14}O_2$, the metabolites remain, in the most part, to be identified (9). Certain metabolites of carbaryl, as formed in vivo by rabbits and goats, are identical with those produced by liver microsomal systems, and include derivatives formed by hydroxylation of the N-methyl group and the 4- or 5-position of the naphthyl group, as well as the 5,6-dihydro-5,6-dihydroxy analog (4, 5). Not any of these carbaryl metabolites are more active as anticholinesterase agents than carbaryl itself (4, 5). Insecticidal and/or anticholinesterase metabolites are formed from other carbamates after injection of 1-dimethyl-carbamoyl-5-methyl-3-pyrazolyl dimethylcarbamate (dimetilan) into insects (10), after injection of 4-methylthio-3,5-xylyl methylcarbamate (Mesurol), 4-dimethylamino-3-cresyl methylcarbamate (Matacil), and 4-dimethylamino-3,5-xylyl

60

methylcarbamate (Zectran) into bean plants (11, 12), and on incubation of certain methyl- and dimethylcarbamates with rat liver microsomal enzymes, based on a failure to lose anticholinesterase activity while forming metabolites which yield formaldehyde on acid degradation (7).

A series of experiments were made with each of 10 carbamate-carbonyl-C^{14}-labeled methyl- and dimethylcarbamates to determine their metabolic fate in a system containing rat liver microsomes and reduced nicotinamide adenine dinucleotide phosphate ($NADPH_2$). With emphasis on those with a carbamate structure, the ether-extractable metabolites were separated by thin-layer chromatography (TLC) and the radioactive spots were located by radioautography. The anticholinesterase activity of the metabolites derived from five methylcarbamates was determined, in situ, on the TLC plates. In some cases, the metabolites were tentatively identified by cochromatography with known compounds.

Methods and Materials

Each of the compounds listed in Table I was utilized with a carbamate-carbonyl-C^{14} label, and each had a radiochemical purity greater than 99 per cent. In addition to $NADPH_2$ and its oxidized form (NADP), nicotinamide adenine dinucleotide in both the oxidized (NAD) and reduced ($NADH_2$) forms was used as a cofactor.

Rat liver was homogenized in 0.25M sucrose and the homogenate was fractionated. The "microsome plus soluble" fraction was the supernatant from centrifugation of the homogenate for 30 minutes at 15,000 g; the "microsome" fraction was the sediment obtained from the microsome plus soluble fraction by centrifugation of the supernatant for an additional 30 minutes at 95,000 g. The microsomal pellet was washed twice with sucrose solution and reconstituted with 0.25M sucrose to give a concentration equivalent to the original 20 per cent homogenate.

In the incubation step, each 25-ml. Erlenmeyer flask contained the following: 2 μmoles of labeled substrate; microsome, or microsome plus soluble, fraction equivalent to 200 mg. of liver; not any or 2 μmoles of cofactor; 250 μmoles of sucrose; 50 μmoles of sodium phosphate; water to make to 2 ml. total volume with a pH of 7.4. The flasks were incubated in air, with shaking, for 4 hours at 37°C. Analysis consisted of ether extraction to recover the original carbamate and carbamate metabolites, and TLC (silica gel G from Kensington Scientific Corp., Berkeley, Calif.; 20 x 20 cm. plates of 0.25 mm. thickness) to resolve these materials. Development of the chromatograms in the case of methylcarbamates involved chloroform-acetonitrile mixture (4 to 1) in the first direction and, after solvent evaporation, ether-hexane mixture (4 to 1) in the other direction;

one-dimensional chromatography with ethyl acetate-ethanol mixture (98 to 2) was used for the dimethylcarbamates. The total radiocarbon present in each of the following fractions was determined by scintillation counting and expressed on a percentage basis: the incubation mixture after incubation but before extraction; the ether and water phases following extraction; each resolved product from TLC, as detected by radioautography.

Cochromatography was used for tentative metabolite identification. The ether extract, containing labeled metabolites, was mixed with 10 to 20 μg, of each non-labeled known compound, the mixture was spotted on a TLC plate, and the plate was developed with the solvent systems as mentioned above, or with ether-carbon tetrachloride mixture (5 to 1). The colored spots for the known carbamates, as detected with ninhydrin or other reagents (4, 10, 13), were compared as to position and shape with darkened areas on the radioautograph, produced by the radiolabeled metabolites.

For localization and assay of cholinesterase inhibitors, the TLC plate containing the resolved radioactive metabolites was sprayed with undiluted human plasma until the gel was wet and appeared glossy. A filter paper (Whatman No. 1) dipped in a plasma-dye mixture was then laid over the wet silica gel, taking care to avoid the trapping of air bubbles. (The plasma-dye mixture consisted of 0.2 per cent (w./v.) water-soluble cresol red in 0.025$\underline{N}$ aqueous sodium hydroxide, to which was added, just before use, an equal volume of undiluted plasma.) After 30 minutes at 27°C., the paper on the plate was sprayed with a 4.5 per cent (w./v.) aqueous solution of acetylcholine bromide until the paper was wet; the plate was then placed in a closed moist chamber at 27°C. Within 30 minutes, red spots appeared on a yellow background when cholinesterase inhibitors were present; these colored spots were compared, as to location, with the darkened areas on the radioautograph. Ether (spectro quality reagent) extracts of incubation mixtures containing the microsomes but not any added cofactor, where little metabolism occurred, were compared to corresponding extracts from incubation mixtures containing microsomes plus $NADPH_2$, where more extensive metabolism was always evident; thus, the "microsomes alone" sample served as a control for the "microsomes plus $NADPH_2$" sample. The results were expressed as the minimum amount of compound in micrograms, based on radioactivity, needed to yield a spot due to the inhibition of pseudocholinesterase.

The methods and materials are described in greater detail in reference number 13.

Results

Table I lists the number of carbamate metabolites, both

TABLE I

Metabolites of Methyl- and Dimethylcarbamate Insecticide Chemicals
Formed by Rat Liver Microsomal Enzymes in Greater than One Per Cent Yield

Insecticide Chemicals	Metabolites Tentatively Identified[1]	Unidentified Metabolites
1-Naphthyl methylcarbamate (carbaryl)	1-naphthyl N-hydroxymethylcarbamate[a] 4-hydroxy-1-naphthyl methylcarbamate[b] 5-hydroxy-1-naphthyl methylcarbamate[a] 5,6-dihydro-5,6-dihydroxy-1-naphthyl methylcarbamate[2]	none
2-Isopropoxyphenyl methyl-carbamate (Baygon)	2-isopropoxyphenyl N-hydroxymethyl-carbamate[a] 2-hydroxyphenyl methylcarbamate[a]	two
3-Isopropylphenyl methyl-carbamate (UC 10854)	none	five
3,5-Diisopropylphenyl methylcarbamate (HRS-1422)	none	six
2-Chloro-4,5-xylyl methyl-carbamate (Banol)	2-chloro-4,5-xylyl N-hydroxymethyl-carbamate[a]	two
4-Methylthio-3,5-xylyl methylcarbamate (Mesurol)	4-methylsulfinyl-3,5-xylyl methyl-carbamate[c] 4-methylsulfonyl-3,5-xylyl methyl-carbamate[c],[3]	one
4-Dimethylamino-3-cresyl methylcarbamate (Matacil)	4-dimethylamino-3-cresyl N-hydroxymethylcarbamate[a] 4-methylamino-3-cresyl methyl-carbamate[d] 4-amino-3-cresyl methylcarbamate[d],[3]	none
4-Dimethylamino-3,5-xylyl methylcarbamate (Zectran)	4-dimethylamino-3,5-xylyl N-hydroxymethylcarbamate[a] 4-methylformamido-3,5-xylyl methyl-carbamate[d] 4-methylamino-3,5-xylyl methyl-carbamate[d] 4-amino-3,5-xylyl methylcarbamate[d]	three
1-Dimethylcarbamoyl-5-methyl-3-pyrazolyl dimethyl-carbamate (dimetilan)	1-methylcarbamoyl-5-methyl-3-pyrazolyl dimethylcarbamate[e],[3]	two
1-Isopropyl-3-methyl-5-pyrazolyl dimethylcarbamate (Isolan)	none	one

1) Sources for known compounds: a) M. H. Balba, Division of Entomology and Acarology, University of California, Berkeley; b) J. B. Knaak, Mellon Institute, Pittsburgh, Pa.; c) C. A. Anderson, Chemagro Corp., Kansas City, Mo.; d) A. M. Abdel-Wahab, Division of Entomology and Acarology, University of California, Berkeley and/or T. R. Norton, Dow Chemical Co., Midland, Mich.; e) E. Knusli, J. R. Geigy S. A., Basle, Switzerland.

2) Previously reported (5).

3) Metabolite yield less than 1.0 per cent.

unidentified and tentatively identified, formed in greater than one per cent yield by the microsomes plus NADPH$_2$ system from 10 methyl- and dimethylcarbamate insecticide chemicals. All of the 10 chemicals gave at least one carbamate metabolite, and half of them gave 4 or more carbamate metabolites. The table does not list metabolites formed by hydrolysis at the carbamic ester site nor those at the origin of the TLC plates.

The extent of metabolism of the carbonyl-C^{14}-labeled carbamates generally increased in the following order, based on the components of the liver microsomal system: 1) microsomes alone; 2) microsomes plus soluble alone; 3) microsomes fortified with NADH$_2$; 4) microsomes plus soluble fortified with NAD; 5) microsomes fortified with NADPH$_2$; 6) microsomes plus soluble fortified with NADP. The substrate-specificity for the microsomes fortified with NADPH$_2$ and the microsomes plus soluble fortified with NADP systems fell in the following categories, based on recovery of the original compound: extensively degraded - Banol, Zectran and Isolan; intermediate stability - carbaryl, Mesurol and dimetilan; most stable - Baygon, UC 10854, HRS-1422, and Matacil. Hydrolysis of the parent compound or its carbamate metabolites was greatest with Isolan (26%), Banol (19%), carbaryl (8%) and Mesurol (8%); water-soluble products were greatest with Banol (46%), Mesurol (43%) and Isolan (43%).

The number of metabolites found to be more potent than the original compound as pseudocholinesterase inhibitors was as follows: carbaryl - not any; Baygon - two unidentified carbamate analogs; Banol - one unidentified carbamate analog; Mesurol - the sulfoxide analog; Zectran - the 4-methylamino and 4-amino analogs.

Discussion

The methyl- and dimethylcarbamate groups are frequently more resistant to metabolism, by microsomal enzymes, than are other groupings in the molecule. Groupings found to be susceptible to oxidation or hydroxylation by microsomal enzymes are as follows:

1. N-Methyl, which is converted to N-hydroxymethyl (carbaryl, Baygon, Banol, Matacil, Zectran), N-formamide (Zectran), or is demethylated (Matacil, Zectran, dimetilan);

2. O-Alkyl, which is dealkylated (Baygon isopropoxy group);

3. S-Alkyl, which is converted to sulfoxide and sulfone analogs (Mesurol);

4. Aromatic ring, which is hydroxylated (carbaryl).

The formation of the additional, unidentified carbamate metabolites can be most easily explained on the basis of hydroxylation at different sites on the ring as well as on ring substituents. There is not any evidence for N-hydroxylation or N-demethylation of the methylcarbamate grouping in Banol, confirming a similar previous study with carbaryl and Baygon (4).

The finding of potent anticholinesterase agents among the metabolites of carbamate insecticide chemicals does not necessarily indicate that these metabolites contribute to the insecticidal activity or to the toxicity to mammals, or that they constitute a hazard as potential residues. This is particularly true of the present findings because pseudo-cholinesterase of plasma rather than true cholinesterase of nervous tissue was used in the assay of metabolites. In addition, it is not known whether these active metabolites are formed and/or persist under in vivo situations in mammals.

Acknowledgment

This study was supported in part by grants from the following sources: the U. S. Public Health Service, National Institutes of Health (Grant No. GM-12248); the U. S. Atomic Energy Commission (Contract No. AT(11-1)-34, Project Agreement No. 113); Supplement 74 to the cooperative agreement between the U. S. Forest Service and the Regents of the University of California; Union Carbide Chemicals Co., New York, N. Y.; Chemagro Corp., Kansas City, Mo.; The Upjohn

Co., Kalamazoo, Mich. The authors are indebted to Louis

Lykken and James Gillett for invaluable assistance and suggestions.

References

1. R. T. WILLIAMS, Detoxication Mechanisms, 796 pp. (1959), Wiley, New York.
2. J. R. GILLETTE, Prog. in Drug Res. 6, 11 (1963)
3. J. E. CASIDA, Radiation and Radioisotopes Applied to Insects of Agricultural Importance, p 223, (1963), Int. Atomic Energy Agency, Vienna.
4. H. W. DOROUGH and J. E. CASIDA, J. Agr. Food Chem. 12, 294 (1964)
5. N. C. LEELING, Ph. D. dissertation, University of Wisconsin, Madison (1965)
6. J. B. KNAAK, M. J. TALLANT, W. J. BARTLEY and L. J. SULLIVAN, J. Agr. Food Chem. 13, 537 (1965)
7. E. HODGSON and J. E. CASIDA, Biochem. Pharmacol. 8, 179 (1961)
8. H.-W. RAHN, Arch. exper. Path. Pharmakol. 241, 157 (1961)
9. J-G. KRISHNA and J. E. CASIDA, J. Agr. Food Chem., in press (1966)
10. M. Y. ZUBAIRI and J. E. CASIDA, J. Econ. Entomol. 58, 403 (1965)
11. A. M. ABDEL-WAHAB, R. J. KUHR and J. E. CASIDA, J. Agr. Food Chem., in press (1966)
12. A. M. ABDEL-WAHAB, Ph. D. dissertation, University of California, Berkeley (1965)
13. E. S. OONNITHAN, Ph. D. dissertation, University of California, Berkeley (1966)

Gas Chromatographic Measurement of Toxaphene in Milk, Fat, Blood, and Alfalfa Hay

by T. E. ARCHER and D. G. CROSBY

Agricultural Toxicology and Residue Research Laboratory
University of California, Davis, California

The need for a sensitive and reproducible method for determination of submicrogram quantities of toxaphene in food, forage
crops, milk, and meat has become apparent with the widespread
use of this pesticide. Graupner and Dunn (1) originally described a spectrophotometric procedure that could determine toxaphene in the 20 to 100 μg range. Extensive cleanup procedures
were employed before analysis.

Kovacs (2) and Faucheux (3) used thin-layer chromatographic
techniques for the rapid detection of micro quantities of several chlorinated pesticide residues. Cleanup of the samples
was of critical importance, and toxaphene streaked on the TLC
plates with the solvent systems employed. Burke and Guiffrida
(4) investigated electron capture gas chromatography for the
analysis of multiple chlorinated pesticide residues in vegetables;
vigorous cleanup procedures were required, and, under their conditions, ten peaks were obtained for toxaphene. Bevenue and
Beckman (5) gas chromatographed toxaphene in the nanogram range
and concluded that in the presence of other chlorinated hydrocarbon pesticides, irrespective of the type of column or column
packing used, no characteristic fingerprint for toxaphene could
be obtained with one possible exception.

Contrary to the comments of these latter authors, the method
described below for the analysis of toxaphene utilized a simple
alkali treatment for cleanup and partial dehydrohalogenation of
the substance, and gas chromatography with an electron capture
detector permitted quantitative analysis. Nanogram quantities
of toxaphene were detected reproducibly.

70

Experimental

Chemicals and Equipment.   Toxaphene was supplied by the manufacturer, all other chemicals were reagent grade, and the solvents were redistilled before use.   The gas chromatograph was an Aerograph Model 200 (Wilkens Instrument Co.) equipped with an electron capture detector and a Texas Instruments Inc. Servo/ Riter II 1 mv. potentiometric recorder.   The chromatographic column was 9' x 1/8" stainless steel packed with 60/80 mesh Chromosorb W (HMDS treated) coated with 5% Dow 710 silicone oil and 5% SE-30 gum rubber.   The 12" section of the column at the injection port was packed with 20/30 mesh calcium carbide for removal of traces of water and alcohol entrained during cleanup.   Nitrogen carrier gas (40 p.s.i., 40-60 ml./min.), a column temperature of 200°C, a detector temperature of 190°C, an injector temperature of 250°C, and an instrument attentuation and sensitivity of 1X was used.   The recorder chart speed was 15"/hour, and the peak areas were measured with a polar planimeter for quantitation.

The infrared spectra were obtained on a Perkin-Elmer Model 221 spectrophotometer, samples being mixed into potassium bromide and analyzed as micro pellets.   Microgram amounts of toxaphene were dehydrohalogenated, extracted into benzene, and concentrated on the potassium bromide prior to pellet preparation. Background interference, found to be insignificant, was determined on a benzene blank treated in a manner similar to the toxaphene solution.

The dehydrohalogenation reagent was prepared fresh for each sample by dissolving 5 g. of C.P. potassium hydroxide in 3 ml. of distilled water followed by the addition of 17 ml. of ethanol.

Analysis of alfalfa hay, milk, fat, and blood.   The preparation and analysis of the samples for toxaphene were performed as described by Crosby and Archer (6) with the exception that recoveries of toxaphene were higher if benzene rather than pentane was used for extraction.

71

Results and Discussion

Five different batches of toxaphene (Hercules Powder Co., Wilmington, Delaware) were analyzed with reproducible results. Fig. 1A represents the gas chromatogram from 30 ng. of untreated toxaphene; the total retention time was approximately 17.5 min. in several poorly resolved peaks. Fig. 1B represents the gas chromatogram of 30 ng. of toxaphene after dehydrohalogenation; the total retention time was approximately 11.0 min. with the major peak at 3.50 min. DDE and related compounds exhibited a longer relative retention time (6.25 min.) than that of the major toxaphene peak. For analytical data, either the measurement of the area of the entire trace from the dehydrohalogenated sample or measurement of the area of the major peak at 3.50 min. was easily accomplished with a planimeter.

Fig. 2 presents the infrared spectra of dehydrohalogenated toxaphene (I) and untreated toxaphene (II). As expected, comparison revealed a decrease in the number of the carbon-chlorine bands in the 12-15 μ range and an increase in the carbon-hydrogen absorption at 3.5 μ. A new carbonyl absorption was evident at 5.8 μ.

Numerous samples of alfalfa hay, milk, bovine fat, and blood have been analyzed by this method in our Laboratory. Background for injections representing as much as 50 mg. of prepared sample was negligible, while recoveries ranged from approximately 74% to 95% when the samples were fortified at the 0.1 ppm. and 0.5 ppm. level. Table I lists recoveries for a few of these samples.

The advantage of this procedure for toxaphene analysis is the very rapid and effective cleanup of the feedstuff and animal products by alkali treatment as well as the chemical conversion of the pesticide into derivatives that are more readily gas chromatographed than the parent substance. The reproducible single peak at 3.50 min. serves for quantitative and qualitative identification, and this peak cannot be confused with those of the DDT group also commonly present in most samples. An additional advantage is that the dehydrohalogenation procedure increased the sensitivity of the electron capture detector to toxaphene by approximately two fold.

We are indebted to Nels Larsen and Eugene Whitehead for technical assistance and to Paul Allen and Joseph Thomas for assistance with the infrared data.

TABLE I

Recovery of Toxaphene in Animal Products and Feedstuffs

Product	Fortification p.p.m.	Recovery %
Alfalfa hay	0.5	95.2
Cow milk	0.1	74.1
Cow milk	0.5	92.7
Cow body fat	0.5	82.2
Cow blood	0.5	88.8
Rat blood	0.5	91.2

References

1. A. J. GRAUPNER and C. L. DUNN, J. Agr. Food Chem. 4, 286 (1960)

2. M. F. KOVACS, JR., J. Assoc. Offic. Agr. Chem. 48, 1018 (1965)

3. L. J. FAUCHEUX, JR., J. Assoc. Offic. Agr. Chem. 48, 955 (1965)

4. J. BURKE and L. GUIFFRIDA, J. Assoc. Offic. Agr. Chem. 47, 326 (1964)

5. A. BEVENUE and H. BECKMAN, Bull. Environ. Contam. and Tox. 1, 1 (1966)

6. D. G. CROSBY and T. E. ARCHER, Bull. Environ. Contam. and Tox. 1, 16 (1966)

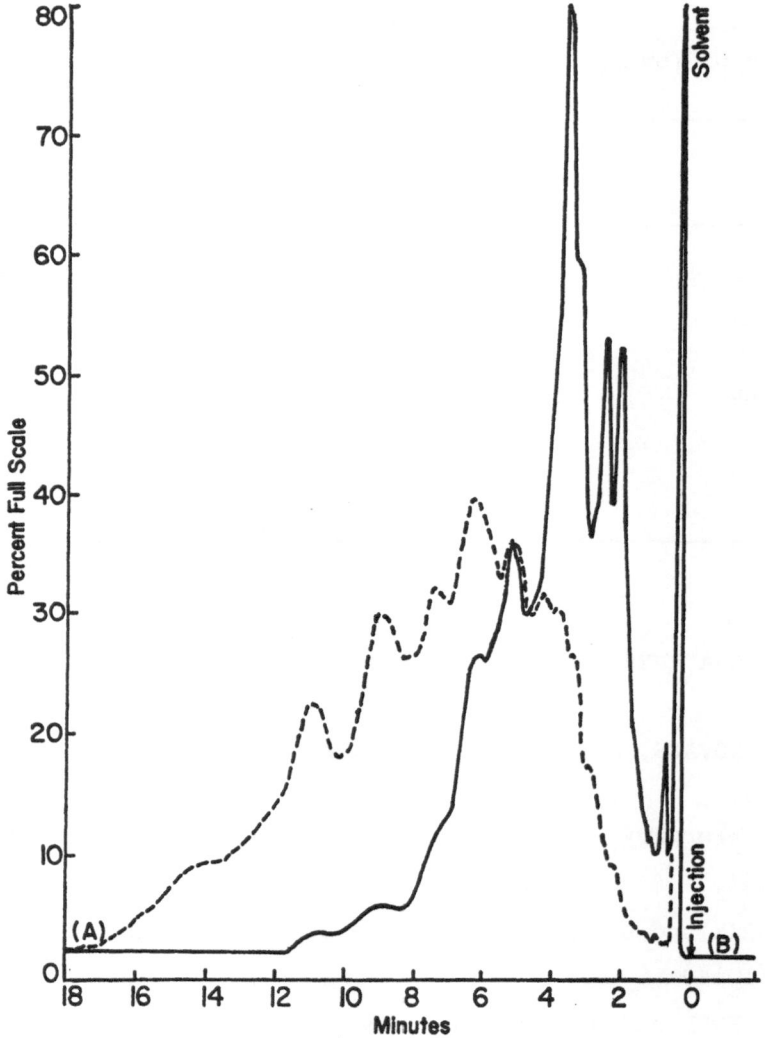

Fig. 1. Gas chromatogram from 30 ng. of toxaphene (A) before alkali treatment and after alkali treatment (B).

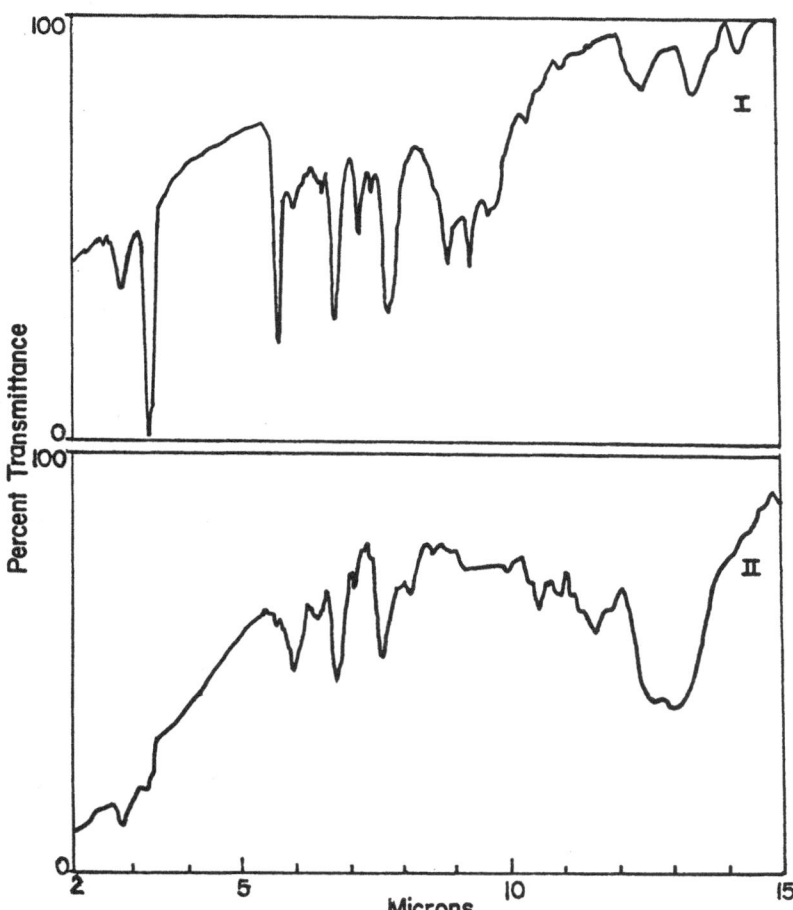

Fig. 2. Infrared spectra of toxaphene after alkali treatment (I) and before alkali treatment (II).

Thin Layer Chromatography of
Dithiocarbamate Fungicides

by J. W. HYLIN
Department of Agricultural Biochemistry
University of Hawaii, Honolulu, Hawaii

During the course of studies on the transformation of dithio-

carbamate fungicides on leaf surfaces, it became necessary to develop

a procedure which would distinguish the residue of the applied

substance from its transformation products. Thin layer chromato-

graphy proved to be the most effective method for this separation

and the following procedure was developed.

Absorbent. Silica-gel G was applied to glass plates in layers

250 microns thick using conventional techniques.

Sample Preparation. Intact leaves were extracted with chloro-

form for 30 minutes on a mechanical shaker. The chloroform extract

was filtered and concentrated to a small volume on a rotating

evaporator. An aliquot of the concentrated extract was applied to

the thin layer plate. Highly pigmented extracts were decolorized

with a minimum amount of Darco G-60 activated charcoal.

Development. Benzene was the solvent used for ascending de-

velopment of plates containing dimethyldithiocarbamates while

benzene-methanol-acetic acid (48:8:4) was used for samples which

contained ethylene-bis-dithiocarbamates.

Published with the approval of the Director, Hawaii Agricultural
Experiment Station as Technical Note No.

Bulletin of Environmental Contamination & Toxicology,
Vol. 1, No. 2, 1966, published by Springer-Verlag New York Inc.

<u>Visualization</u>. The compounds of interest were located after development using the following spray reagent.(1)

a) 1.5 Gm cupric chloride and 3 gm ammonium chloride dissolved in 50 ml water containing 3 ml of concentrated ammonia.

b) 20 Gm hydroxylamine hydrochloride dissolved in 100 ml of water. Equal volumes of a) and b) are mixed just prior to use.

Ziram,thiram,tetramethylthiuram monosulfide, and Zineb give bright yellow spots gradually changing to green, while Maneb yields a brown spot. The lower limit of detectability is approximately 2.5 micrograms.

TABLE

Rf values for dithiocarbamates

	Rf	
	S_1*	S_2
Thiram	.17$\pm$.02	.86$\pm$.02
Tetramethylthiuram monosulfide	.30$\pm$.02	-
Ziram	.68$\pm$.02	.94$\pm$.02
Maneb	.98$\pm$.02	.98$\pm$.02
Zineb	.95$\pm$.02	.88$\pm$.03

* S_1 = Benzene, S_2 = Benzene-Methanol-Acetic Acid (48:8:4)

REFERENCE

1.F.FEIGL, Spot Tests in Organic Analysis, p 233(1956), Elsevier.

Residue Reviews

Residues of Pesticides and Other Foreign Chemicals in Foods and Feed

Edited by FRANCIS A. GUNTH

University of California, Rive with an Advisory B representing eleven coun

Residues of pesticides and othe eign" chemicals in foodstuffs are of concern to one everywhere; they are essential to food proc and manufacture, yet without surveillance and gent control some of those that persist could a conceivably endanger the public health.

The object of *Residue Reviews* is to provic cise, critical reviews of timely advances, philo and significant areas of accomplished or need deavor in the total field of residues of these che in foods, in feeds, and in transformed food pr

The scope of *Residue Reviews* is internatio encompasses those matters, in any country, wh involved in allowing pesticide and other pla tecting chemicals to be used safely in producin ing, and shipping crops.

Bulletin

Contents

Bulletin of Environmental Contamination and Toxicology

AIMS AND SCOPE

The Bulletin of Environmental Contamination and Toxicology will provide rapid publication of significant advances and discoveries in the fields of pesticide residue research, air, soil, and water contamination and pollution, methodology, and other disciplines concerned with the introduction, presence, and effects of toxicants in the total environment.

Results of current research will be presented as brief reports providing information which is potentially useful to all individuals concerned with environmental contamination.

The articles will be free from restrictions imposed by purely scientific journals, particularly with respect to completeness of the studies reported and the attendant delays in publication.

Descriptions of new methods, procedures, or techniques shall be sufficiently detailed so as to permit direct application in other laboratories.

Review articles and obvious abstracts of papers forthcoming in other publications are not invited and probably will not be acceptable.

Articles suitable for inclusion shall be relatively short (less than 2,000 words) and will be prepared following specific instructions to permit reproduction by the photo-offset process from the original manuscript.

It is the hope of the Editorial Board that this Bulletin will provide a meeting ground for researchers who daily encounter problems related to the contamination of our environment and who welcome opportunities to share in new discoveries as they occur.

The Bulletin will be issued six times a year. This will be raised to 12 issues annually as demand increases.

Published bi-monthly by SPRINGER-VERLAG NEW YORK INC., 175 Fifth Avenue, New York, N. Y. 10010, Telephone (212) 673-9797.

Separation and Detection of Submicrogram Quantities of Pesticides by an Improved TLC Technique

by HERMAN BECKMAN and WRAY WINTERLIN
Agricultural Toxicology and Residue Research Laboratory
University of California, Davis, California

In the last few years thin-layer chromatography has developed into one of the most important techniques available for the separation and identification of components of a mixture. It has paralleled gas chromatography in importance and period of development.

The spreading and diffusion of spots, particularly at increased Rf values is often a problem with thin-layer and paper chromatography. This paper describes an approach to a procedure in which the developing spots are prevented from spreading; thereby permitting a more sensitive means of detection.

In 1962, Gamp, et. al. (2) described a procedure using ribbed glass plates for thin-layer chromatography. Such plates are not always available and they suffer from the disadvantage of being difficult to prepare with standard equipment. Uneven layers in each groove make solvent migrations uneven and

Presented at the 150th National Meeting, American Chemical Society, September, 1965.

78

comparisons difficult. Adjustment of the width of the strip is not possible with the fixed rib plates.

We have chosen to call the new procedure "thin-strip thin-layer chromatography" (TSTLC). The idea is to prepare plates composed of thin parallel strips of adsorbent. Each strip then functions as a separate thin-layer column or plate, yet all of the strips are equal in resolving power, as they are all developed at the same time under the same conditions.

A group of organo-phosphorus pesticides was chosen to demonstrate the technique and to show that small quantities were detectable as compared to a standard plate. An alternate procedure that will produce a similar plate, but with much thicker strips can be followed. This will provide hemi-cylindrical columns of adsorbent that are capable of carrying greater quantities of materials to be separated, yet retaining the characteristics of the previously described plates.

Experimental

Several approaches were made in the development of the final procedure described. A process now being considered is to use a cylindrical piece such as a section of large bore glass tubing in place of the glass plate. The thin strips may then be applied vertically or in a spiral up the cylinder in order to lengthen the column. A piece of tubing eight inches long by 2.5 inches diameter would provide opportunity for as many strips as an 8 x 8 inch plate, but longer strips if applied as a spiral.

The thin-layer plates (8 x 8 inch) were prepared from Silica-Gel H, by mixing 30 grams of the dry powder with 33 ml. of 0.1 N hydrochloric acid and 33 ml. of absolute methanol and applying the material to the glass plates with a Desaga-Brinkmann applicator set for a thickness of 250 microns. The plates were allowed to air dry for 15 minutes followed by oven drying at 110^0 C for 45 minutes.

The plates were stripped by means of a modified window

cleaner as illustrated in Fig. 1.

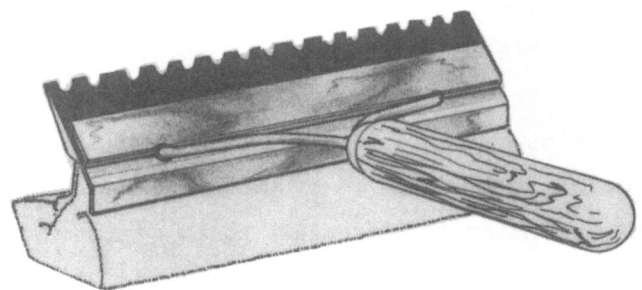

Fig. 1. Tool for preparing TSTLC
chromatographic plates.

The tool was made from an 8 inch rubber squeegee, notched with
a small circular file. The notches may be made by freezing the
rubber at sub-zero temperatures in a dry ice-acetone bath and
notching with a file upon immediate withdrawal from the bath.

By resting one edge of the plate and stripper against a smooth
object such as a board, a smooth even stroke of the stripper
drawn across the plate results in uniform and straight strips.
Fig. 2 presents a demonstration of the appearance of plates with
developed chromatograms, comparing a standard TLC plate with
a TSTLC plate.

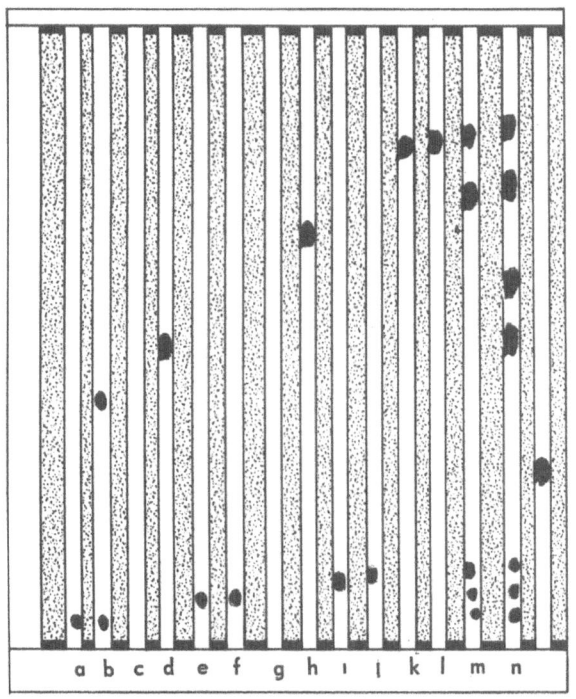

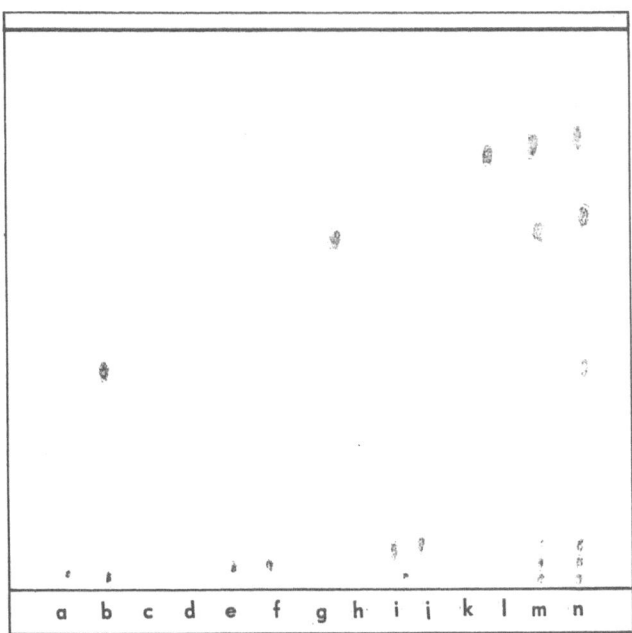

Fig. 2. Appearance of a TSTLC and a standard TLC plates with developed chromatograms. Identity of the compounds is given in the text and Table I.

A TSTLC plate and standard plate which was not stripped were each spotted with 1 and 2.5 microgram each of Systox, Parathion, Guthion, Thimet, Imidan, and Trithion respectively. The plates were then developed in a solvent mixture of 10 per cent n-hexane in toluene. Using the ascending technique the solvent front was permitted to travel 16 cm. from the original spotting. The plates were then dried by a stream of warm air and the separated spots were detected by spraying with a solution of 0.5 per cent 2,6-dibromoquinone-chlorimide w/v in methanol (3). After heating in an oven for 10 minutes at 70° C. the yellow to brown spots were easily detectable on the white background depending, of course, on the concentration of the specific pesticide. Table 1 presents the Rf values for each of these materials.

TABLE 1

Rf values for the pesticides used to demonstrate the method as described in the text and shown in Fig. 1.

Pesticide	Strips identified by letters as shown in figure 1	Rf
Systox	a, b	0.44
Parathion	c, d	0.53
Guthion	e, f	0.04
Thimet	g, h	0.73
Imidan	i, j	0.09
Trithion	k, l	0.88

An alternate procedure for TSTLC was also utilized by spreading and stripping while the Silica Gel was wet. This technique was performed by spreading a heavy layer of the Silica Gel slurry at one end of a plate and, by utilizing the stripper mentioned previously, immediately stroking across the plate with a smooth even stroke. The plate may be restroked two or three times to give a more consistent and even strip from one end of the plate to the other.

Results and Discussion

In Fig. 3, two types of TSTLC plates are illustrated by a cross-section view of the thin strips.

CROSS-SECTIONAL VIEW OF TSTLC

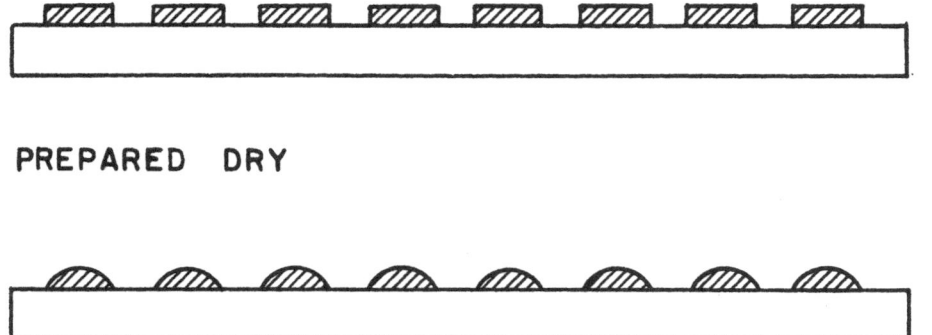

PREPARED DRY

PREPARED WET

Fig. 3. Cross-section view of the glass plates showing the appearance of the silica gel strips prepared on a dry plate and as they would appear when prepared from a wet slurry.

On the left, the strips are flat since they were formed after the plate was dried, and on the right is the cross-section view of the plate made when the Silica Gel was wet. The wet slurry procedure requires more refinement in technique as the ratio

of solution to adsorbent is critical to the point where over wetting will cause the strips to flatten out and insufficient wetting
will cause the strips to be lumpy and not uniform. A degree
of flexibility in the width of strips prepared may be obtained
by using two equally grooved rubber plates that may be adjusted to control the opening for the size of the strip.

These micro strips on plates allow the separation and detection of submicrogram quantities of the compounds described
earlier. The sensitivity of detection is the same regardless
of Rf value as the narrow strips do not permit the spreading of
the spots as they progress up the plate. The size of the strips
can be adjusted to the desire of the investigator both in width
and in thickness. However, if the strips are too narrow, the
band of the developed area lengthens somewhat. This vertical
spread will depend, to a large degree, on the type of compounds worked with as well as the amount of material to be
chromatographed.

TSTLC permits many analyses to be carried out on a single
plate depending, of course, on the width of the strips. In our
investigation we found that strips 4 mm. wide were optimum
and as many as 20 strips could be developed on one 8 inch x 8
inch plate. Extraction or removal of separated spots is more
easily accomplished using TSTLC. One may spot and develop
two strips in the same way using one for color development and
the other for recovery of the compound at the corresponding Rf.

The compounds we selected for illustration in this paper
were six difficult-to-separate organophosphorus pesticides.
With the TSTLC technique and solvent system mentioned, we
were able to separate all six organophosphorus compounds. An
isomer of Systox was also separated.

A recent study involving the isolation and identification of
products of parathion photodecomposition has allowed us to
combine the use of this TSTLC technique and the sodium

thermionic detector. Since the phosphorus sensitive detector on the gas chromatograph will respond to nanogram quantities of these compounds it is very useful in such a study. The use of larger amounts of compounds will allow identification of separated spots by infrared spectroscopy. This three-way detection system should aid greatly in the identification of the products being sought. The STD equipment in use is a conversion of a Wilkins Hi-Fy chromatograph and was described at a recent symposium on pesticide analysis. (1)

Literature Cited

1. Beckman, H. and Gauer, W. O., "Studies on the Function and Operating Parameters of the Sodium Thermionic Detector". Presented at the Wilkins Instrument and Research, Inc., Symposium on Pesticide Residue Analysis, June, 1965.

2. Gamp, A., Studer, P., Linde, H., and Meyer, K., Experiementia (Basil), 18, 292 (1962).

3. Stahl, E. ed. "Thin-Layer Chromatography A Laboratory Handbook", p. 489, 1965. Springer-Verlag, N.Y.

Persistence of Methyl Parathion Residues on Sunflower Seeds[1]

by H. W. Dorough, N. M. Randolph, and G. H. Wimbish
Department of Entomology
Texas A&M University, College Station, Texas

Growers of minor crops, or crops that yield marginal profits, face a particular problem in their efforts to control insect pests. Potential insecticide usage is often not great enough to stimulate widespread interest in securing data necessary to obtain a label for its use on the crop. Sunflower is a crop that falls within this category.

The sunflower moth, Homoeosoma electellum (Hulst.) is a major pest of sunflower. Damage is caused by the larva feeding in the maturing seedhead and thereby yields are drastically reduced. Performance tests have indicated that methyl parathion could be used to control this pest provided residues on the seeds at time of harvest were not excessive.

Experimental

Methyl parathion, diluted to deliver 5 gallons of spray per acre was applied by use of a high clearance spray machine to sunflower plants at rates of 0.5 and 1.0 pounds per acre. Treatments were made 28, 21, 14, 7, 3, and 1 day before the predetermined harvest date. Samples were collected immediately after the final treatment (0-day applications) by random cutting of 15 sunflower heads from each treated and check plot. Seeds were threshed from the heads, placed in plastic bags and held in a deep freeze until analyzed.

Precipitation on the third, eighteenth and twenty-first day following the initial application was 0.89, 1.40 and 1.45 inches, respectively.

Methyl parathion was extracted from the seeds by blending 100-gram samples in 300 ml. of acetonitrile, filtering the homogenate and concentrating the filtrate to about 10 ml. The residue was taken up in 50 ml. of benzene and dried over anhydrous sodium sulfate before adding the extract to a chromatographic column. The column, 2 x 30 cm., was prepared by adding successive layers of the following components: (1) 12 grams of acid alumina, Brockmann Activity 1, 80-200 mesh; (2) five grams of a 1:1 mixture of Hyflo Super-Cel and Darco G-60 carbon; (3) 12 grams of acid alumina. Methyl parathion was eluted from the column with 180 ml. of benzene and the eluate concentrated to a suitable volume, minimum of one ml., for injection of aliquots into a gas chromatograph.

Gas chromatographic analyses were made with a Barber Colman Series 5000 instrument equipped with a sodium thermionic detector (1). A six foot, 1/4 inch i.d. glass column packed with 10 per cent DC 200 silicone fluid on 80-90 mesh Anakrom ABS was used. Operating temperatures were as follows: column 215° C, injector 240° C, and detector 220°C. Nitrogen was used as the carrier gas at 60 ml./min. and air was supplied to the burner at 300 ml./min. Hydrogen flow was adjusted to maintain a baseline current of 5×10^{-9} amp. The sensitivity scale was set at 10^{-8} AFS and the recorder chart speed at 15 inches/hour. The retention time for methyl parathion was 4.75 minutes.

Results and Discussion

Following extraction and clean-up, an equivalent of 100 mg. of untreated seeds injected into the gas chromatograph gave no peaks with retention times similar to methyl parathion. When methyl para-thion was added to 100 grams of the seeds at the 0.10 ppm level, recovery of the insecticide ranged from 94 to 102 per cent. Based on a peak height of one cm. for two nanograms of material, the sensitivity of the method was 0.02 ppm.

Analysis of treated sunflower seeds revealed that relatively low deposits were present immediately following treatment (Table 1); only 2 ppm methyl

TABLE 1

Levels of Methyl Parathion in Sunflower Seeds Following
Treatments at 0.5 and 1.0 Pounds Per Acre.

| Days After Treat-ment | P.P.M. After Treatment at | | | | | |
| | 0.5 lbs/a | | | 1.0 lbs/a | | |
	Max.	Min.	Ave.[2]	Max.	Min.	Ave.[2]
0[1]	0.800	0.541	0.696	2.120	1.737	2.068
1	0.318	0.278	0.304	0.933	0.846	0.970
3	0.114	0.076	0.100	0.490	0.352	0.401
7	0.093	0.040	0.060	0.197	0.135	0.184
14	0.023	0.019	0.023	0.095	0.086	0.089
21	ND[3]	ND	ND	0.068	0.053	0.059
28	ND	ND	ND	ND	ND	ND

[1] Samples collected immediately after treatment.
[2] Average of four analyses.
[3] None detectable.

88

parathion residues were found on the seeds from plants treated at the higher application rate. Sunflower seedheads grow in such a manner that the seeds face the ground and are protected in umbrella fashion from above. The low insecticide deposits on the seeds probably resulted because the seeds were shielded from the downward spray. Such distribution of the toxicant on the plant would be highly desirable since the sunflower moth larva enters the heads at the stem end and would be subjected to much higher residues than present on the seeds.

Dissipation rates of the methyl parathion residues were similar following treatment at the two different levels. In both cases, over 50 per cent of the initial deposits were lost from the seeds within 24 hours. Based on the 1.0 ppm methyl parathion residue that is presently allowed on many agricultural crops, it was evident that dissipation was sufficiently rapid to allow harvest of seeds three days after treatment at one pound per acre. Residues of this magnitude were not deposited when plants were treated at 0.5 pounds per acre. If allowable residues were set as low as 0.10 ppm, the maximum interval between methyl parathion treatments of 0.5 and 1.0 pounds per acre and harvest would be seven and 14 days, respectively, as indicated by the results of this study.

References

(1) L. Giuffrida, JAOAC <u>47</u>, 293(1964).

Automated Determination of Orthophosphate:
An Application Designed Especially for the Combusted
Product of Organophosphorous Pesticide Residues

by Daniel E. Ott and Francis A. Gunther
Department of Entomology
University of California, Riverside, California

The present method, of interest to residue chemists who are
employing the BLINN (1) or other Schöniger flask combustion method
for determining organophosphorus pesticide residues, will remove
much of the tedium of the analysis as well as improve the precision.
It should also apply to the orthophosphate combusted product of
organophosphorus pesticide residues no matter how this product is
obtained, i.e., from combustion tube furnace, Parr bomb, or wet
oxidation techniques.

Method

Apparatus and Reagents

(a) AutoAnalyzer system which is an adaptation of that of
WEINSTEIN et al. (2) arranged as shown in Figure 1 (equipment
available from Technicon Controls, Inc., Ardsley, N. Y.).

(b) Ammonium molybdate reagent.--3% (w./w.) in water.

(c) ANSA reagent.--Dissolve 150 g. of sodium bisulfite and
5.0 g. of sodium sulfite in about 800 ml. of water, heat to about
50° C., and add 2.5 g. of 1-amino-2-naphthol-4-sulfonic acid. Stir
until nearly dissolved, dilute to 1,250 ml., filter, and store this

Bulletin of Environmental Contamination & Toxicology,
Vol. 1, No. 3, 1966, published by Springer-Verlag New York Inc.

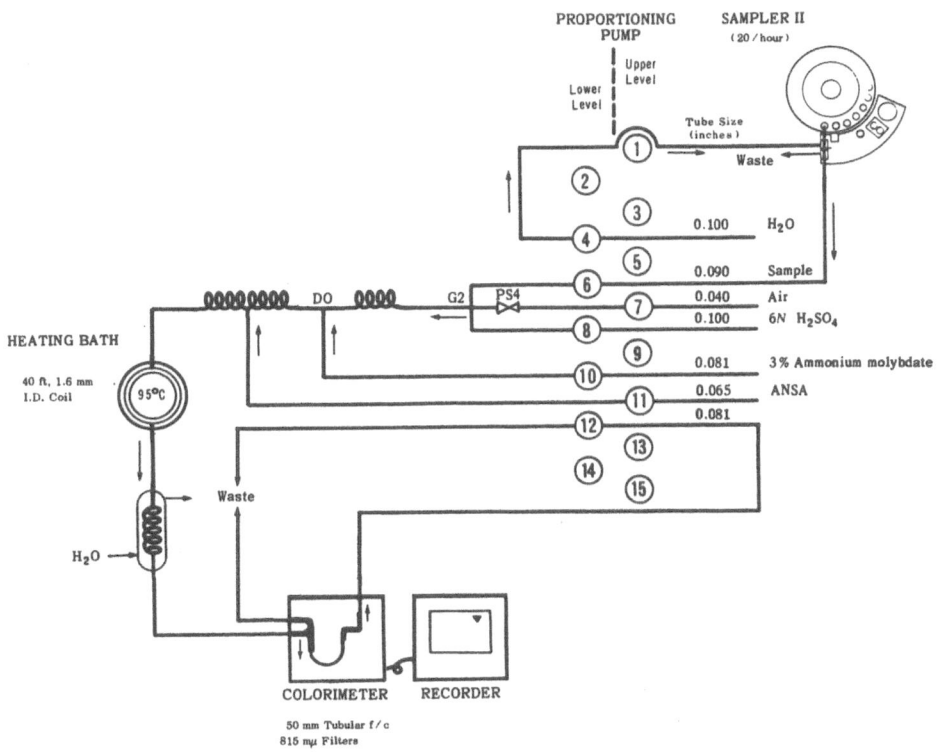

Fig. 1. AutoAnalyzer system for the determination
of orthophosphate

stock solution in an amber bottle in the refrigerator; dilute 200

ml. to one liter for a working solution.

(d)   Phosphorus standard solutions.--Dissolve an appropriate

amount of KH_2PO_4 in 0.75$\underline{N}$ sulfuric acid and dilute to one liter

with the same. Prepare working solutions by dilution of this

stock solution with 0.75$\underline{N}$ sulfuric acid.

(e)   Lever IV.--(Technicon Chemical Company, Inc., Ardsley,

N. Y.). Just before use, add 0.5 ml. of this detergent per

liter to each reagent which is pumped into the analytical system.

Procedure

Standards.--Prepare a calibration curve for each pesticide to be examined by sampling orthophosphate standard solutions over the concentration range which would be obtained from the complete combustion or other oxidation of the levels of pesticide to be analyzed. Pour about 5-ml. portions of these solutions into standard macro sample cups for the Sampler II module and operate it in a 20-samples-per-hour mode with a cam providing the same time (1.5 minutes) for wash between samples as for sampling.

Samples (recommended).--Cleanup and combust in a Schöniger flask a 100-g. crop sample according to the procedures of BLINN (1). About 5 ml. of the 0.75$\underline{N}$ sulfuric acid solution which has been used to absorb the combustion products in the Schöniger flask can be transferred directly to a sample cup and analyzed as above for "Standards."

Results and Discussion

Figure 2 presents data from standard solutions obtained in this manner to show the type of calibration curve reasonably obtainable from 100-g. samples of a crop fortified at the indicated levels of Guthion, parathion, and dimethoate, respectively, if one assumes 100% recoveries in the cleanup and combustion procedures.

For routine analyses similar calibration curves, but plotted directly on a Technicon universal chart reader, should be prepared from fortified control samples for both the highest and lowest percentage phosphorus species of organophosphorus insecticides likely to be encountered in the crop to be analyzed. Then each unknown

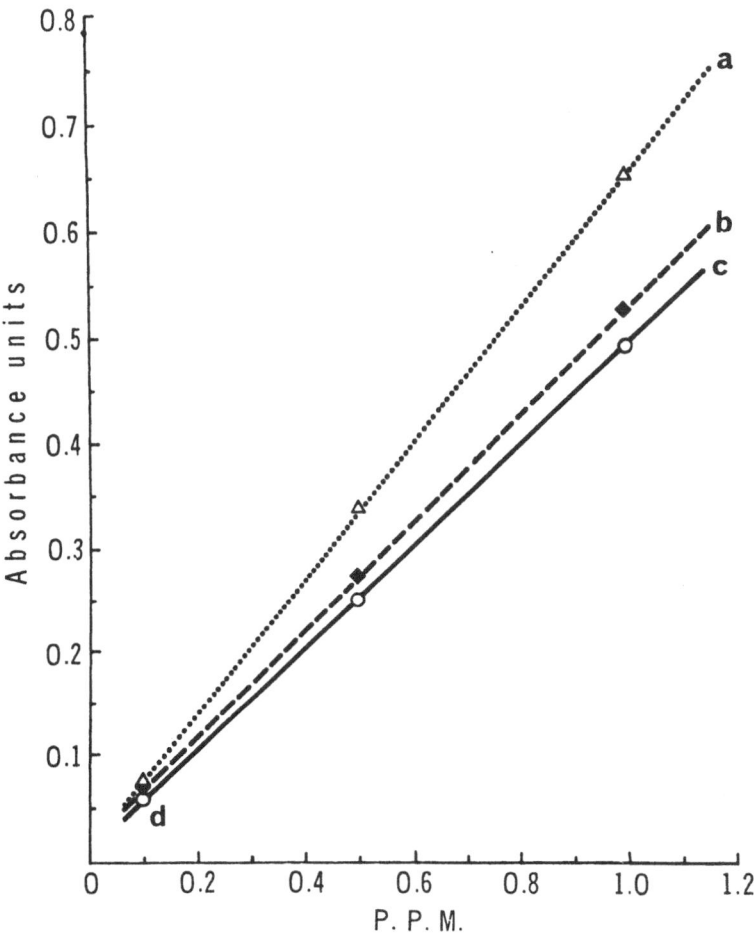

Fig. 2. Standard curves of orthophosphate in
0.75N sulfuric acid: levels which would be obtained from p.p.m. fortifications of dimethoate
(a), parathion (b), and Guthion (c), respectively,
in 100 g. of crop, assuming 100% recoveries in
cleanup and combustion techniques; (d) the lowest point is equivalent to 0.1 μg. P/ml. and was
replicated five times at 0.055 ± 0.001 absorbance
unit (see Fig. 3)

sample peak, when the chart paper is pulled under the chart reader,

can be read off quickly in terms of p.p.m. of these two insecticides;

all other organophosphorus insecticides likely present in the crop

would fall between the two levels read out.

An indication of the precision obtainable with this analytical

system is found in Figure 3. Note the close agreement between replicates and especially between replicates of low-level standards before
and after high-level standards. Figure 3 also shows the appearance
of a typical Technicon chart reader standard curve superimposed on
the chart record. These results are from a non-expanded recorder

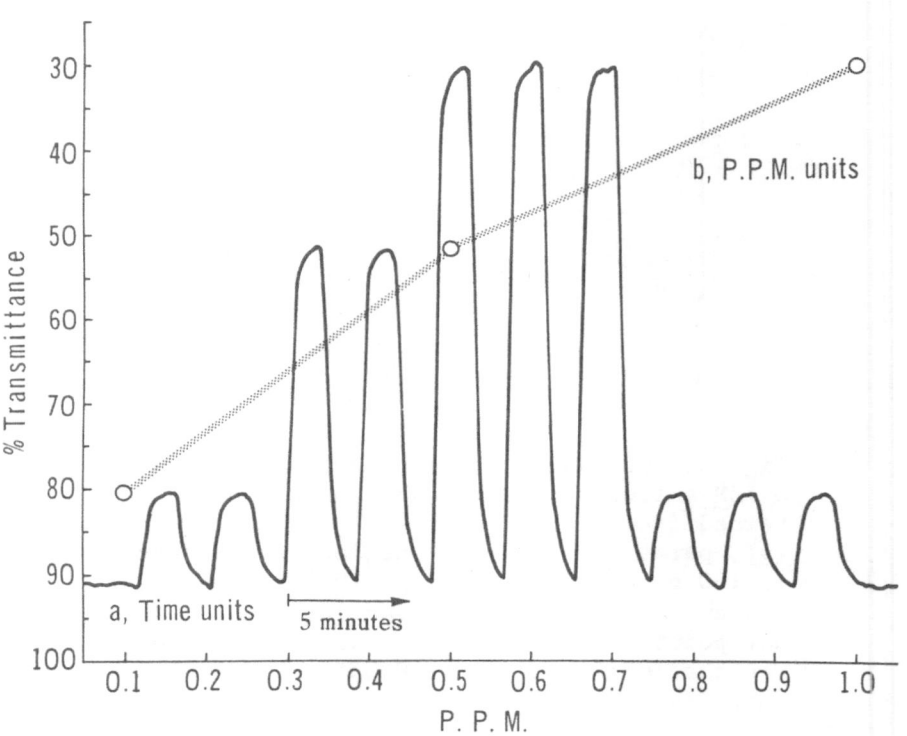

Fig. 3. Typical chart recording and superimposed Technicon chart reader standard curve obtained from standard solutions of orthophosphate
in 0.75N sulfuric acid; standard curve expressed
in terms of p.p.m. of Guthion from 100 g. of
crop as in Figure 2

range; if greater sensitivity is required a Range Expander (Technicon) can be conveniently employed.

If this system is employed with samples of characteristics grossly different from those recommended here, as for example solution in alkali rather than in sulfuric acid, the concentration of diluent sulfuric acid (see Fig. 1) may need to be adjusted from 6$\underline{N}$ to some new concentration to produce optimum colorimetric response.

A forthcoming report (3) will describe the present system employed with an automated continuous wet digestion-oxidation system for converting organophosphorus insecticides to orthophosphate.

Summary

An automated colorimetric analytical determination of ortho-phosphate is reported. The method should be especially adaptable to the combusted products from a Schöniger flask combustion technique but is adaptable to other combustion procedures as well. The analytical procedure is rapid, convenient, precise, and responds readily to well below 0.1 µg. of phosphorus, as orthophosphate, per ml. in the final analytical solution.

Acknowledgments

Technical assistance of Carol A. Lazzaro is gratefully acknowledged. Financial and other assistance were generously provided by the Edanros Research Foundation, Inc., New York, N. Y. Paper No. 1672, University of California Citrus Research Center and Agricultural Experiment Station, Riverside, Calif.

References

(1) R. C. BLINN, J. Agr. Food Chem. 12, 337 (1964).

(2) L. H. WEINSTEIN, R. F. BOZARTH, C. A. PORTER, R. H. MANDL, and
     B. G. TWEEDY, Contrib. Boyce Thompson Inst. 22, 389 (1964).

(3) D. E. OTT and F. A. GUNTHER, In preparation for J. Assoc.
     Official Agr. Chemists.

The Effects of Combinations of Insecticides on Susceptible and Resistant Mosquito Fish[1]

by Denzel E. Ferguson and C. Rex Bingham
Department of Zoology, Mississippi State University
State College, Mississippi

Although the effects of two pesticides acting simultaneously upon non-target organisms has received little attention in laboratory investigations, animals living near certain types of croplands may be exposed to two or more insecticides regularly. In cotton-producing areas of the Mississippi Delta, most fields receive 10-15 insecticide applications a season, usually between June 1 and September 10. The most commonly used insecticides are endrin, DDT, toxaphene, and methyl parathion, applied either individually, in some pattern of rotation, and/or mixed in a variety of combinations. Adjacent fields may be on different spraying schedules and treated with different insecticides or insecticide combinations. Most insecticides are applied with airplanes, especially after midsummer when the cotton plants get tall. Where drift and runoff from such operations reach nearby bodies of water, mosquito fish (<u>Gambusia affinis</u>) and several other species of fish may exhibit high levels of resistance to a variety of pesticides (1, 2, 3, 4, 5).

---

[1] Supported by grant ES 00086-01 from the Office of Resource Development, U. S. Public Health Service.

97

Bulletin of Environmental Contamination & Toxicology,
Vol. 1, No.3, 1966, published by Springer-Verlag New York Inc.

The present account compares the responses of a highly resistant population of mosquito fish from a heavily contaminated site and a susceptible population from an uncontaminated locality to simultaneous exposures to all possible paired combinations of endrin, DDT, toxaphene, and methyl parathion.

Materials and Methods

Resistant mosquito fish were obtained from a ditch that drains and bisects several large cotton fields near Belzoni, Mississippi. Susceptible fish were collected from a pond on non-agricultural land near State College, Mississippi. All fish were collected with a fine-meshed seine and held overnight in the laboratory prior to testing.

Technical grade samples of the four insecticides, provided by the manufacturers, were prepared as 1% solutions in acetone and diluted to desired test concentration with tapwater (pH 7.8, hardness 28 ppm). In the initial dilution, additional acetone was employed to facilitate dissolution of the insecticides, but the amount was regulated so as not to exceed 2 ml per liter of final test solution.

In one series of tests, samples of 50 mosquito fish were exposed for 36 hours in 20 liters of insecticide solution in 15-gallon aquaria. Four aquaria were required to test fish from one population against a pair of insecticides, e.g., an aquarium containing insecticide solution A, an aquarium containing insecticide solution B, an aquarium containing a mixture of A

and B at the same concentrations placed in the separate aquaria,

and an aquarium of tapwater for controls. In this manner,

samples from both susceptible and resistant populations were

tested against endrin-DDT, endrin-toxaphene, endrin-methyl

parathion, DDT-toxaphene, DDT-methyl parathion, and toxaphene-

methyl parathion. Because of differences in tolerance of the

two populations, it was necessary to use higher insecticide

concentrations in tests of resistant fish, especially for endrin

and toxaphene (Figs. 1 & 2). Mortality was recorded at 15- or

30-minute intervals early in the tests, followed by hourly

observations or 6 hour checks, depending upon the progress of

the test.

All tests were run at temperatures of 72 ± 4F, and fish were

not fed.

Results and Discussion

In the aquarium tests, control mortality never exceeded one

fish out of 50. In the treatments, the much higher

concentrations used for the resistant population resulted in less

mortality than was produced by lower concentrations used in tests

of susceptible fish (Figs. 1 & 2). Whereas the combination of

two insecticides produced higher mortality among resistant fish

than did the individual insecticides, the combination scarcely

exceeded the individual kills of toxaphene and endrin in the

tests of susceptible fish. The relative positions of the

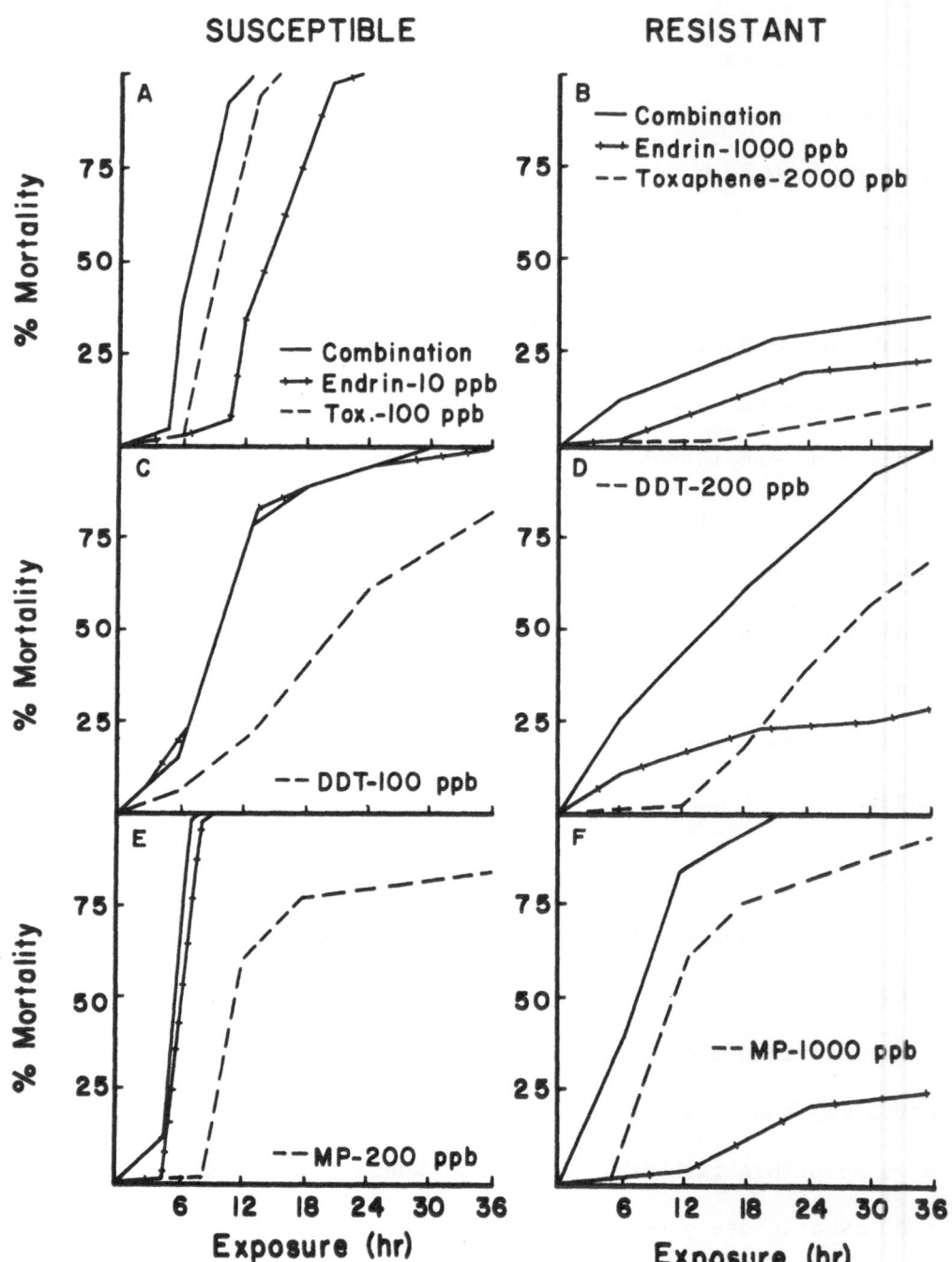

Figure 1. Mortality produced in samples of 50 mosquito fish
from resistant and susceptible populations by two
insecticides separately and in combination.

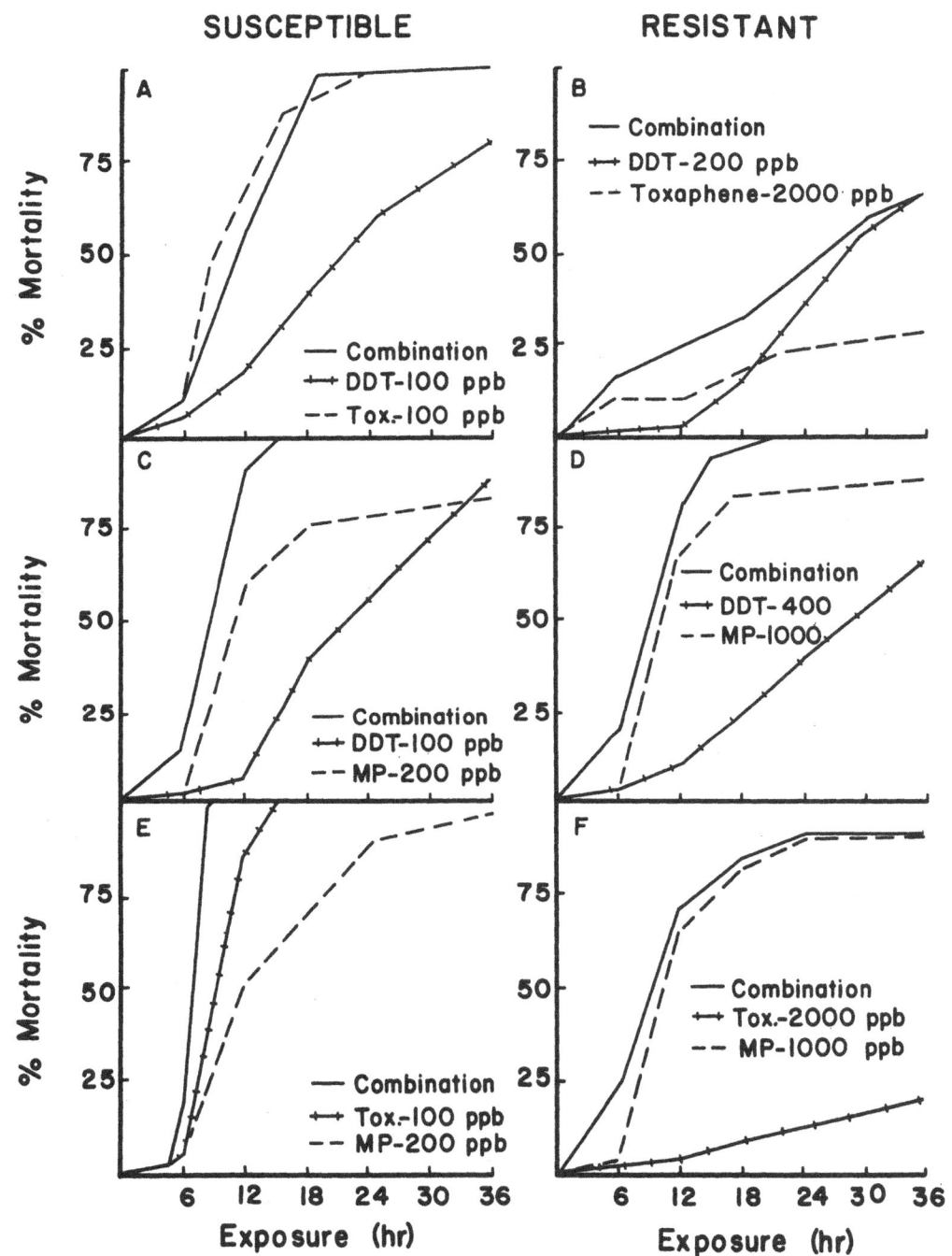

SUSCEPTIBLE

RESISTANT

A
% Mortality
Combination
DDT-100 ppb
Tox.-100 ppb

B
Combination
DDT-200 ppb
Toxaphene-2000 ppb

C
% Mortality
Combination
DDT-100 ppb
MP-200 ppb

D
Combination
DDT-400
MP-1000

E
% Mortality
Combination
Tox.-100 ppb
MP-200 ppb

F
Combination
Tox.-2000 ppb
MP-1000 ppb

Exposure (hr)

Exposure (hr)

Figure 2.   Mortality produced in samples of 50 mosquito fish
from resistant and susceptible populations by two
insecticides separately and in combination.

101

mortality curves for the individual insecticides (compared with the combination curve) are reversed in equivalent tests of susceptible and resistant populations, except in the DDT-methyl parathion tests (Fig. 2, C & D). In general, the results reflect the extreme levels of endrin and toxaphene resistance in the resistant population--i.e., endrin and toxaphene are relatively more toxic to the susceptible fish; DDT and methyl parathion are relatively more toxic to the resistant fish. The mortality curve for methyl parathion shows a consistent deflection after 12 hours, probably indicating rapid hydrolysis of the compound. The results failed to indicate additive effects wherein the combination mortality exceeded the sum of the mortalities produced by the individual insecticides.

The results of the jar tests agree with the findings of the tests run in aquaria, i.e., the combination of two insecticides produced higher mortality than did either insecticide alone. In all tests, the sum of the mortalities caused by individual insecticides exceeded that for the same insecticides in combination, hence, there were no additive effects. In tests of resistant fish involving the endrin-toxaphene, endrin-methyl parathion, the toxaphene-methyl parathion combinations, the mixture produced levels of mortality very close to the total mortality produced by the compounds separately. In general, higher mortality from mixtures probably indicates differences in the modes of action of the toxicants involved.

Literature Cited

1. S. B. VINSON, C. E. BOYD, and D. E. FERGUSON, Science 139, 217 (1963).

2. C. E. BOYD and D. E. FERGUSON, Mosquito News 24, 19 (1964).

3. C. E. BOYD and D. E. FERGUSON, J. Econ. Entomol. 57, 430 (1964).

4. D. E. FERGUSON and C. E. BOYD, Copeia 1964, 706 (1964).

5. D. E. FERGUSON, D. D. CULLEY, W. D. COTTON, and R. P. DODDS, BioScience 14, 43 (1964).

Determination of 2,4-D Residues in Animal Products

by D. G. Crosby and J. B. Bowers

Agricultural Toxicology and Residue Research Laboratory
University of California, Davis, California

Several simplified methods for the analysis of pesticide residues in animal products have been reported recently from our Laboratory (1). The present communication is concerned with a rapid and convenient method for cleanup and analysis of high protein samples for the herbicide 2,4-dichlorophenoxyacetic acid (2,4-D).

Marquardt and Luce (2) applied a colorimetric analysis to 2,4-D in milk, and Coakley, et al. (3), used the same procedure for determination of the acid and its esters in shellfish and fish. Burchfield and Storrs (4) and Marquardt, Burchfield, and Storrs (5) have described a method for milk analysis based on microcoulometric gas chromatography using internal standards. In each of these instances, the isolation of 2,4-D in a form suitable for analysis proved to be rather complicated, and each analytical method presented serious special problems. The procedure described here, in which milk was employed as a typical high-protein substrate, proved to be simple, sensitive, and relatively inexpensive.

Experimental

Chemicals and Equipment. All chemicals were reagent grade, 2,4-D was a recrystallized analytical standard, and the reagent grade solvents were redistilled shortly before use. Diazomethane in ether was prepared from "Diazald" according to directions of the manufacturer (Aldrich Chemical Co., Inc., Milwaukee, Wisconsin). The gas chromatograph was an Aerograph Model 600B (Wilkens Instrument Co.) equipped with an electron capture detector. A 5' x 1/8" stainless steel column packed with either 5% Dow 11 silicone oil or 5% SE-30 gum rubber on 60/80 mesh Chromosorb W was satisfactory. A column temperature

104

of 197°C, injection port temperature of 240°C, and nitrogen carrier gas flow of 50 ml./min. at 14 p.s.i. were maintained.

Cleanup. A sample of milk was warmed to 40°C, mixed well, and a 25 ml. subsample was transferred to a 250 ml. centrifuge bottle. Petroleum ether (25 ml.), diethyl ether (25 ml.), and 1 N aqueous sodium hydroxide solution (1 ml.) were added, the stoppered bottle was shaken mechanically for 30 min. to extract fats and other neutral substances, and the layers were caused to separate by centrifugation for 15 min. at 1700 r.p.m. The upper layer was decanted, and the ether-petroleum ether extraction was repeated. The aqueous phase then was transferred to a 300 ml. flask fitted with a reflux condenser, and 25 ml. of methanol and 2 g. of sodium hydroxide were added, the solution was boiled for one hour, cooled, acidified with 250 ml. of 0.5 N hydrochloric acid, and the 2,4-D was extracted into three successive 15 ml. portions of chloroform. At this point, the extract may be pure enough for analysis; if not, it may be washed with two 25 ml. portions of 1% aqueous sodium bicarbonate solution, the combined bicarbonate washes acidified, and the released 2,4-D reextracted into chloroform.

Analysis. An aliquot (40%) of the chloroform extract was evaporated under a stream of air, 1 ml. of an ether solution of diazomethane was added, and, after about 5 min., the solution again was evaporated. Hexane (1.0 ml.) and 6% aqueous sodium sulfate solution (5 ml.) were added, and an aliquot was subjected to gas chromatography. One microliter of the hexane solution represented 10 mg. of milk and contained one ng. of 2,4-D (as the methyl ester) if the original milk level was 0.1 p.p.m.

Results and Discussion

As shown in Table 1, recoveries of 2,4-D standards were approximately 90%, and the practical limit of detection was 0.5 ng. in a sample representing 10 mg. of milk (0.05 p.p.m.). Retention time of 2,4-D methyl ester on the SE-30 column was 5 min.; under the same conditions, aldrin and DDE had retention times of 15 min. and 40 min., respectively, so that the desired peak was well-separated from other common pesticide contaminants.

TABLE 1

Recovery of 2,4-D From Milk

Concentration Calculated (p.p.m.)	Concentration Found (p.p.m.)	Quantity Found (ng.)	Recovery %
0	0	0	-
0.050	0.042	0.42	84
0.10	0.093	0.93	93
0.20	0.18	1.80	90

Although intended primarily for the estimation of 2,4-D, this procedure should apply equally well for the analysis of other phenoxy herbicides such as 2,4,5-T, 2,4-DB, Silvex, and MCPA; halogenated phenylacetic acids such as Fenac; and other halogenated acids including trichlorobenzoic acid and Dicamba. The procedure is suitable for the free acids and their salts; esters and amides, removed in the initial ether-petroleum ether extraction, undoubtedly would require much more detailed clean-up before they could be analyzed in this way.

It has long been observed that high-protein samples appear to bind 2,4-D, and the methods developed to avoid this difficulty have become quite elaborate. Recoveries may be low, and, in our experience, high electron-capture backgrounds may be encountered. The present method conveniently and effectively avoids these problems and appears to offer general application to blood, meat, and many other animal products.

References

1. D. G. CROSBY and T. E. ARCHER, Bull. Environ. Contam. and Toxicol. 1, 16 (1966).

2. R. P. MARQUARDT and E. N. LUCE, Anal. Chem., 23, 1484 (1951).

3. J. E. COAKLEY, J. E. CAMPBELL, and E. F. MCFARREN, J. Agr. Food Chem., 12, 262 (1964).

4. H. P. BURCHFIELD and E. E. STORRS, Abstracts 140th National Meeting, American Chemical Society, September, 1961, p. 19A.

5. R. P. MARQUARDT, H. P. BURCHFIELD, E. E. STORRS and A. BEVENUE. In "Analytical Methods for Pesticides, Plant Growth Regulators, and Food Additives," G. Zweig, Ed., Academic Press, New York, 1964. Vol. IV, p. 95.

Effect of 2,4-Dichlorophenoxyacetic Acid on Different Metabolic Types of Bacteria

by Lewis T. Hart and A. D. Larson
Department of Bacteriology and Agricultural Experiment Station
Louisiana State University, Baton Rouge, Louisiana

The effect of auxins (plant hormones) on microorganisms has been studied by several investigators. The bacteriostatic effects of 2, 4-dichlorophenoxyacetic acid (2, 4-D), a growth-regulating substance, was first demonstrated by Stevenson and Mitchell (13) and by Lewis and Hammer (3). Dubos (1) observed that a number of synthetic, unsaturated, ring-containing acids including 2, 4-D, exerted bacteriostatic effects on certain microorganisms.

Worth and McCabe (16) reported that aerobic bacteria are susceptable to inhibition by 2, 4-D but facultative and anaerobic microorganisms are not. Magee and Colmer (11), employing manometric procedures, found that the respiration of old cells of Azotobacter was more sensitive to 2, 4-D than were young cells.

108

The work of Heath and Clark (4, 5) and Johnson and Colmer (7, 8, 9, 10) concerning the mode of action of auxins and related compounds on micro-organisms suggested interference with metal ion metabolism. Johnson and Colmer (7) demonstrated that the inhibition of oxygen utilization in Azotobacter vinelandii and Rhizobium meliloti by 2, 4-D was related to the concentration of available magnesium and that the simultaneous addition of phosphate with 2, 4-D enhanced the toxicity of 2, 4-D to A. vinelandii.

The studies reported here were performed to determine the sensitivity of certain metabolic types of bacteria to 2, 4-D and, if possible, to determine if specific enzyme systems are involved.

Materials and Methods

Organisms. Pseudomonas denitrificans ATCC 13867, Pseudo-monas stutzeri, ATCC 11607, and Corynebacterium nephridii, ATCC 11425 (Achromobacter nephridii, Hart et al (2))were obtained from the American Type Culture Collection. Bacillus subtilus, Bacillus megaterium, Corynebacterium faciens, Achromobacter parvulus, Clostridium perfringens, Clostridium acetobutylicum, Clostridium tetanomorphum, Clostridium sporogenes, Escherichia coli, Aerobacter aerogenes, Staphylococcus aureus, and Bacillus cereus were obtained from the departmental culture collection. Aerobic, denitrifying and facultative organisms were maintained on nutrient agar slants. Anaerobic organisms were subcultured in a broth medium composed of 1% peptone, 1. 0% glucose, and 0. 01% sodium thioglycollat

109

Growth experiments. The basal medium at pH 7. 0 was composed of 1. 0% Bacto peptone and 1. 0% glucose. Sodium thioglycollate (0. 01%) was added to the basal medium for the growth of anaerobic organisms. Sodium nitrate (0. 5%) or sodium nitrite (0. 2%) was incorporated into the basal medium for the growth of the denitrifying bacteria in the absence of air. For growth under aerobic conditions, 250 ml Erlenmeyer flasks containing 50 ml of medium were incubated on a New Brunswick rotary shaker at 30°C. Anaerobic conditions were obtained by employing 13 x 100 mm screw cap tubes completely filled with medium. Purified 2, 4-D was obtained from Dow Chemical Company, Midland, Michigan. The sodium salt of 2, 4-D, prepared by adjusting the pH to 7. 0 with NaOH, was incorporated into the growth medium in graded amounts. Growth media were inoculated with a standardized suspension of bacteria (86% transmission at 600 mμ in a Bausch and Lomb Spectronic 20 spectrophotometer); 0. 002 ml of the standardized suspension was used for each ml of medium. Growth of bacteria was measured by determining the change in OD at 600 mμ. Optical density readings were made every 24 hours and the results reported are those after which no changes occurred.

Manometric studies. Cells of the organisms were grown in a broth medium composed of 1. 0% peptone, 1. 0% glucose, and 0. 5% $NaNO_3$, harvested by centrifugation, washed twice with 0. 1 M phosphate buffer, pH 7. 2, and resuspended in the same buffer. Standard manometric procedures as described by Umbriet, Burris, and Stauffer (14) were followed.

Results

Growth experiments with aerobic organisms. All the aerobic organisms tested were quite sensitive to 2, 4-D although different degrees of sensitivity within the group existed (Table 1).

TABLE 1.

Effect of 2, 4-D on the growth of aerobic bacteria

Organism	Incubation time hr	μmoles 2, 4-D/ml					
		0	1	2	3	4	5
B. subtilis	24	1.50*	0.36	0.21	0.10	0.03	0.00
B. megaterium	64	0.68	0.10	0.00	0.00	0.00	0.00
C. faciens	72	1.00	0.54	0.12	0.05	0.03	0.00
A. parvulus	72	0.95	0.94	0.93	0.95	0.95	0.00

* change in OD at 600 mμ.

Growth of B. subtilis, B. megaterium, and C. faciens, all Gram-positive organisms, was markedly reduced by 1.0 μmole/ml of 2, 4-D and no growth occurred at 5.0 μmoles/ml. The growth response of the Gram-positive bacteria to increasing concentrations of 2, 4-D was graded, while the response of the Gram-negative organism, A. parvulus, was an all or none effect. The Gram-negative A. parvulus was less sensitive to 2, 4-D than were any of the four Gram-positive cultures studied.

111

Growth experiments with anaerobic bacteria. Growth of clostridia showed a uniform sensitivity to 2, 4-D (Table 2).

TABLE 2

Effect of 2, 4-D on the growth of anaerobic bacteria

Organisms	Incubation time hr	μmoles 2, 4-D/ml				
		0	1	5	10	15
C. perfringens	24	0.55*	0.55	0.55	0.47	0.00
C. acetobutylicum	24	0.68	0.66	0.61	0.42	0.00
C. tetanomorphum	48	0.70	0.68	0.65	0.46	0.00
C. sporogenes	48	0.70	0.66	0.64	0.53	0.00

* change in OD at 600 mμ

Little effect on growth was observable at a concentration of 10 μmoles/ml and all four species were strongly inhibited by 15 μmoles/ml of 2, 4-D. The clostridia, although Gram-positive, were much less sensitive to 2, 4-D than any of the aerobic bacteria studied.

Growth experiments with facultative organisms. The growth of both Gram-positive and Gram-negative bacteria of this metabolic type was less sensitive to 2, 4-D under aerobic than anaerobic conditions (Table 3). The Gram-positive bacteria were affected by much lower concentrations of 2, 4-D under both aerobic and anaerobic environments than were the Gram-negative organisms. In fact, the Gram-positive

TABLE 3

Effect of 2, 4-D on growth of facultative bacteria

Organism				μmoles 2, 4-D/ml					
	0	1	2	5	10	15	30	40	50
E. coli									
aerobic	0.90*	0.86	0.87	0.85	0.85	0.84	0.64	0.20	0.00
anaerobic	0.50	0.44	0.40	0.30	0.30	0.33	0.11	0.00	0.00
A. aerogenes									
aerobic	1.00	0.96	0.99	1.00	0.83	0.75	0.50	0.03	0.00
anaerobic	0.30	0.31	0.30	0.26	0.21	0.20	0.06	0.00	0.00
S. aureus									
aerobic	0.36	0.20	0.16	0.00	–	–	–	–	–
anaerobic	0.14	0.07	0.00	0.00	–	–	–	–	–
B. cereus									
aerobic	0.70	0.35	0.25	0.09	0.03	0.00	–	–	–
anaerobic	0.13	0.06	0.04	0.00	0.00	0.00	–	–	–

* change in OD at 600 mμ, incubation time of 24 hours
– = no change

bacteria under aerobic conditions were inhibited by lower concentrations of 2, 4-D than anaerobically grown Gram-negative microorganisms.

Growth experiments with denitrifying bacteria. Under aerobic conditions the growth of the three organisms employed was affected less by 2, 4-D than under anaerobic conditions with nitrate as electron acceptor (Table 4). Cells growing under anaerobic conditions with nitrite as electron acceptor displayed a sensitivity to 2, 4-D close to that of aerobically-grown cells. P. denitrificans was less sensitive to 2, 4-D under all growth conditions than were A. nephridii and P. stutzeri which displayed about the same sensitivity to 2, 4-D.

113

TABLE 4

Effect of 2,4-D on growth of denitrifying bacteria

Organisms	Incubation time hr	2,4-D μmoles/ml						
		0	5	10	15	20	30	40
P. denitrificans								
aerobic	60	0.95*	0.95	0.75	0.64	0.50	0.50	0.02
anaerobic (NO_3-)	72	0.41	0.28	0.30	0.21	0.00	0.00	0.00
anaerobic (NO_2-)	72	0.24	0.26	0.27	0.26	0.24	0.12	0.05
P. stutzeri								
aerobic	36	1.00	1.00	0.90	0.28	0.07	0.00	0.00
anaerobic (NO_3-)	36	1.00	0.10	0.04	0.00	0.00	0.00	0.00
anaerobic (NO_2-)	36	0.30	0.25	0.19	0.12	0.16	0.00	0.00
A. nephridii								
aerobic	72	1.00	0.90	0.80	0.70	0.20	0.00	0.00
anaerobic (NO_3-)	72	0.50	0.15	0.00	0.00	0.00	0.00	0.00
anaerobic (NO_2-)	72	0.25	0.25	0.23	0.25	0.06	0.00	0.00

* change in OD at 600 mμ

Manometric experiments. Manometric experiments were performed

to compare the effect of 2,4-D on the reduction of nitrate and nitrite by

resting cells of nitrate-grown P. denitrificans. When nitrate was supplied

as the electron acceptor (Fig. 1), 2,4-D exhibited a marked effect on gas

production. The rate of gas formation was decreased by approximately

50% in the presence of 3.5 μmoles/ml of 2,4-D. The presence of 7.5

μmoles/ml of 2,4-D completely inhibited gas production. Increasing

amounts of 2,4-D resulted in a similar decrease of endogenous gas

production (Fig. 1).

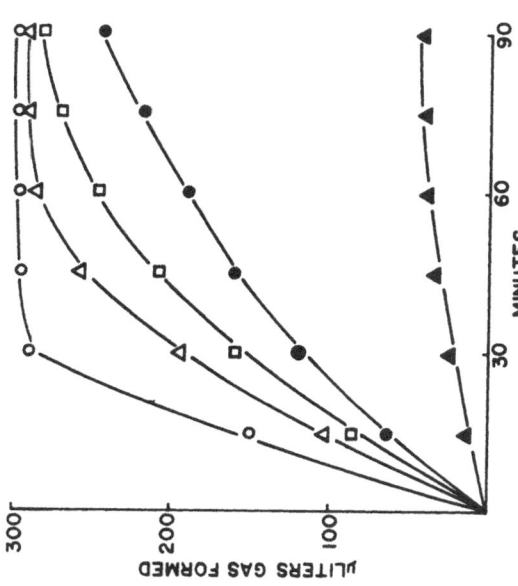

Figure 2. Effect of 2, 4-D on gas production from nitrite by resting cells of P. denitrificans. The reaction vessels were prepared as described in Figure 1 except that NaNO₃ was substituted for NaNO₃. Symbols: ○ = without 2, 4-D; △ = 10 μmoles, 2, 4-D; □ = 20 μmoles, 2, 4-D; ● = 30 μmoles, 2, 4-D; ▲ = endogenous.

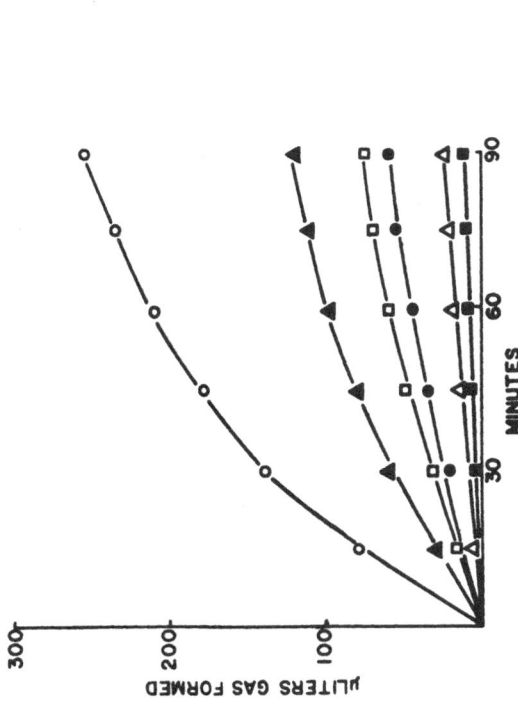

Figure 1. Effect of 2, 4-D on gas production from nitrate by resting cells of P. denitrificans. The reaction vessels contained 0.5 ml of 0.2 M potassium phosphate, pH 7.1, 12 mg of cells, 0.2 ml of 20% KOH in the center well, 25 umoles of NaNO₃, 25 umoles glucose, 2, 4-D and distilled water to a volume of 3.0 ml. Gas phase, N₂; Temperature, 30°C. Symbols: ○ = without 2, 4-D; ▲ = 10 μmoles, 2, 4-D; □ = 20 μmoles, 2, 4-D; ● = endogenous; △ = endogenous with 10 μmoles, 2, 4-D; ■ = endogenous with 20 μmoles, 2, 4-D.

Figure 2 shows the results of manometric experiments utilizing

nitrite as the added electron acceptor. Although the rate of gas formation

was decreased by 2, 4-D, good gas production occurred in the presence

of 10.5 μmoles/ml of 2, 4-D. Endogenous activity from nitrite was not

affected by the amounts of 2, 4-D used in these experiments.

The addition of magnesium or molybdenum at zero time prevented

the inhibitory action of 2, 4-D, but reversal of inhibition could not be

demonstrated by addition at a later time.

Discussion

It is apparent that different bacteria display varied sensitivities

to 2, 4-D and that some correlation exist between the response to 2, 4-D

and the physiology and enzyme content of microorganisms.

Newcombe and Downing (12) reported, but did not present data,

that Gram-positive bacteria are more sensitive to 2, 4-D than Gram-

negative organisms. Our results show that A. parvulus (Gram-negative)

was markedly more sensitive to 2, 4-D than were the Gram-positive

clostridia. Comparisons of sensitivity to the herbicide within the same

metabolic type (within the facultative group or within the aerobic

organisms) demonstrated quite clearly the greater sensitivity of the

Gram-positive bacteria.

Worth and McCabe (16) reported that the growth of aerobic organisms was inhibited by 2, 4-D but that growth of facultative and anaerobic bacteria was not affected by 2, 4-D at concentrations of 2% (80 µmoles/ml when calculated on the basis of pure 2, 4-D). Our results differ in that growth of clostridia was prevented by 15 umoles of 2, 4-D/ml and none of the Gram-negative facultative bacteria grew at concentrations of 40 µmoles/ml under either aerobic or anaerobic conditions. The Gram-positive facultative microorganisms, under both conditions of oxygen availability, were much more sensitive to 2, 4-D than their Gram-negative counterparts.

The variance in results between our work and that of Worth and McCabe (16) may be explained, at least in part, by the differences in experimental design. Worth and McCabe (16) used solid media, different microorganisms, and determined growth visually. They made no comparison in sensitivity of facultative organisms to 2, 4-D under aerobic and anaerobic growth conditions nor were denitrifying bacteria studied.

Consideration of the variation in sensitivity of the bacteria to 2, 4-D provides some insight as to the enzyme or enzyme systems which may be most affected by 2, 4-D. The marked sensitivity of the aerobic as compared to the anaerobic bacteria indicates that the cytochrome system may be particularly sensitive to the herbicide. That this type of reasoning may not be correct is implied from the results obtained with the facultative bacteria. If the cytochrome system were most sensitive to 2, 4-D, then it might be expected that the growth of facultative organisms would be inhibited more strongly under aerobic than anaerobic conditions. The results of our experiments demonstrated that growth of facultative bacteria was equally or somewhat more sensitive to 2, 4-D under anaerobic conditions.

If the electrons flow through the same or similar iron porphyrin systems prior to the reduction of oxygen or nitrate, then the nitrate reductase must be more sensitive to 2, 4-D than is cytochrome oxidase. Nitrite reduction probably occurs without involvement of sensitive iron porphyrin enzymes. It is interesting to note that 2, 4-D had an effect on nitrate endogenous activity similar to that when an exogenous electron donor was present. Since 2, 4-D exerted no observed effect on nitrite endogenous activity, this is further evidence for the greater sensitivity of the nitrate-reducing system as compared to the nitrite-reducing system.

It has been suggested by Williams (15) that 2, 4-D affects bacteria by decreasing permeability. Our results suggest otherwise since the denitrifying bacteria show a difference in response under different growth conditions. Permeability mechanisms would be the same in these bacteria irrespective of the final electron acceptor; therefore, an effect on permeability is not indicated.

It may be noted that our work does not necessarily cast doubt on the theory that, in bacteria, 2, 4-D acts primarily by interfering with mineral metabolism. Our experiments may only denote the enzyme(s) which are most sensitive to metal deprivation. Experimentation with isolated enzymes is one method whereby the mode of action of 2,4-D on microorganisms can be finally elucidated and such experiments are now under way in our laboratory.

Acknowledgment

The senior author is a Trainee under Public Health Service Grant NIH, 5T1-GM692.

References

1. R. J. DUBOS, Proc. Soc. Exptl. Biol. and Med. 63, 317 (1946)

2. L. T. HART, A. D. LARSON and C. S. McCLESKEY, J. Bacteriol. 89, 1104 (1965)

3. R. W. LEWIS and C. L. HAMNER, Mich. Agric. Exptl. Sta. Quart. Bull. 29, 112 (1946)

4. O. V. S. HEATH and J. E. CLARK, Nature 178, 600 (1956)

5. O. V. S. HEATH and J. E. CLARK, Nature 177, 1118 (1956)

6. E. J. JOHNSON and A. R. COLMER, App. Microbiol. 3, 123 (1955)

7. E. J. JOHNSON and A. R. COLMER, J. Bacteriol. 73, 139 (1957)

8. E. J. JOHNSON and A. R. COLMER, J. Bacteriol. 73, 1118 (1957)

9. E. J. JOHNSON and A. R. COLMER, Antibiot. and Chemotherap. VII, 99 (1957)

10. E. J. JOHNSON and A. R. COLMER, Plant Physiol. 33, 99 (1958)

11. L. A. MAGEE and A. R. COLMER, Weeds 4, 124 (1956)

12. A. S. NEWMAN and C. E. DOWNING, J. Agric. Food and Chem. 6, 352 (1958)

13. E. C. STEVENSON and J. W. MITCHELL, Science 101, 642 (1945)

14. W. W. UMBREIT, R. H. BURRIS, and J. F. STAUFFER, Manometric Techniques (1957) Burgess Publishing Co., Minneapolis.

15. V. R. WILLIAMS, J. Bacteriol. 79, 125 (1960)

16. A. W. WORTH JR. and A. N. McCABE, Science 108, 16 (1948)

Editorial

From time to time, the process by which the Bulletin is published will permit a member of the Editorial Board to comment on his favorite subject or on some current need or development in his field of specialization. I would like to take such an opportunity to welcome all our readers, subscribers and authors who together are contributing to the development and growth of the Bulletin. As you read this, Issue 4 will be at the printers and manuscripts for Issue 5 are accumulating. These early issues have offered many interesting articles but of necessity are not truly representative of the full scope which the Bulletin hopes to attain in the future. Our associate editors are working actively in their areas of specialization to provide manuscripts in the diverse subjects which are encompassed by the aims and goals of the Bulletin. Even so it will be some time before we do have the full range of subjects which we hope to cover in this periodical. To speed the process of subject diversification, I would ask each reader to inform his colleagues of the availability of the Bulletin as a publishing medium for subjects in the area covered by our aims and goals. In addition, should any of you have suggestions for improving the Bulletin, I can assure you that the Editorial Board is most receptive and each will give every comment full consideration.

Bulletin

Contents

Bulletin of Environmental Contamination and Toxicology

AIMS AND SCOPE

The Bulletin of Environmental Contamination and Toxicology will provide rapid publication of significant advances and discoveries in the fields of pesticide residue research, air, soil, and water contamination and pollution, methodology, and other disciplines concerned with the introduction, presence, and effects of toxicants in the total environment.

Results of current research will be presented as brief reports providing information which is potentially useful to all individuals concerned with environmental contamination.

The articles will be free from restrictions imposed by purely scientific journals, particularly with respect to completeness of the studies reported and the attendant delays in publication.

Descriptions of new methods, procedures, or techniques shall be sufficiently detailed so as to permit direct application in other laboratories.

Review articles and obvious abstracts of papers forthcoming in other publications are not invited and probably will not be acceptable:

Articles suitable for inclusion shall be relatively short (less than 2,000 words) and will be prepared following specific instructions to permit reproduction by the photo-offset process from the original manuscript.

It is the hope of the Editorial Board that this Bulletin will provide a meeting ground for researchers who daily encounter problems related to the contamination of our environment and who welcome opportunities to share in new discoveries as they occur.

The Bulletin will be issued six times a year. This will be raised to 12 issues annually as demand increases.

Published bi-monthly by SPRINGER-VERLAG NEW YORK INC., 175 Fifth Avenue, New York, N. Y. 10010, Telephone (212) 673-9797. Six issues per year. Subscription price: $15 per year, for institutions, $7.50 per year for individuals.

Apparent Organobromine Compounds in Higher Plants by Neutron-Activation Analysis

by F. A. GUNTHER
Department of Entomology
University of California, Riverside, California

and R. E. SPENGER
Chemistry Department
California State College, Fullerton, California

In a broad search (1) for organochlorine compounds in plant

parts never exposed to pesticides an occasional occurrence in

plants of apparent organobromine compounds was demonstrable

by neutron-activation analysis (2). Finely ground plant parts

were dehydrated with chloride- and bromide-free isopropyl alco-

hol then coextracted with chloride- and bromide-free n-hexane;

the hexane extract was washed free of isopropyl alcohol and of

inorganic halides with chloride- and bromide-free distilled

water, concentrated about 50-to-1, then thermal-neutron activa-

ted and scanned, after a suitable decay interval, for 37-minute

Cl^{38} and for both 18-minute Br^{80} and 36-hour Br^{82} by gamma-ray

spectrometry (2). In Table I are presented typical results

from plants and some of their parts grown on virgin land never

exposed to a pesticide. Some of these data represent crops and

harvests over several years (1).

Bulletin of Environmental Contamination & Toxicology,
Vol. 1, No. 4, 1966, published by Springer-Verlag New York Inc.

Table I.

Apparent organochlorine and organobromine residues in plant parts.

Plant	Plant part	Chlorine p.p.m.[a,b/]	Bromine p.p.m.[a,c/]
Atriplex lentiformis	Leaves and stems	2.4, 3.3	4.40±0.03
			4.90±0.03
Schinus Molle	Tops	1.7, 1.4	1.10±0.02
Suaeda sp.	Whole plants	5.5, 3.3	0.30±0.01
Beets	Tops	0.3	0.15±0.02
Broccoli	Heads	Nil	0.13±0.01
Chard	Foliage	0.2	0.20±0.01
Corn, sweet	Leaves	0.2, 0.3	0.51±0.02
Cucumber	Seeds[d/]	1.7, 1.4[e/]	9.91±0.0!
Barley	Whole plants	---	0.34±0.02
			0.26±0.02
Lima beans	Whole plants	0.9	0.20±0.01
Onions	Whole plants	0.1	0.18±0.01
			0.27±0.05
Peppers	Whole plants	0.4	0.18±0.01
Radishes	Tops	Nil	0.10±0.01
Zucchini	Fruits	0.2, 0.3	0.13±0.01
Several crops	Edible parts	---	~1[f/]

[a/] Duplicate values are from separate samples and usually separate crops. Gamma radiation measurements at 1.65 Mev for Cl and 0.55 Mev for Br. The hexane, isopropyl alcohol, and water used in preparing samples for analysis were free of chlorides and bromides, within the limits of detection by neutron-activation analysis, after 50-to-1 concentration.

[b/] Reference (1).

[c/] Reproducibility from counting statistics.

[d/] Commercial source.

[e/] 555, 605, and 672 p.p.m. Cl by combustion chloride analysis (15) because of the high interference of Br⁻ in the direct potentiometric determination of Cl⁻.

[f/] Reference (3).

Neutron-activation analysis for bromine is highly specific and reproducible (2) so there can be little doubt that this element was actually present in the amounts indicated. Whether it was actually organically bound bromine is another matter in view of the recently demonstrated (1) solubility in hexane, and carry-over through repeated water washes of hexane solutions, of chloride ion as a phosphatidyl choline complex. Despite the lack in Table I of any relationship between chloride and bromide values for specific samples, bromide ion may also complex with these omnipresent lecithin-type substances to give false positives in organobromine-determining pesticide and other residue analyses involving added or metabolized organobromine compounds. Thus, one should be immediately suspicious of all "organobromine" data below about 0.5 p.p.m. (cf. Table I). On the other hand, except possibly for some _Senecio_ spp. there are apparently no naturally occurring organochlorine compounds elaborated by higher plants (4), but there are claimed some plant-elaborated organobromine and organofluorine compounds and numerous organoiodine compounds, as illustrated in Table II.

The only really positive organohalides in higher plants would be the fluorine compounds in the two _Dichapetalum_ spp. The report of bromoacetic acid in wine is rather old and somewhat doubtful because this chemical has sometimes been used as a wine "preservative". The organoiodine compound(s) reported in wheat do not seem to have been very well characterized as yet, and even with jaconine some authors feel that the iodine present is an arti-

fact of the isolation. Thus, there are organochlorine and organobromine compounds from the microorganisms and bromine and iodine
compounds from the more complex thallophytes as the only really
positively identified naturally occurring organohalogen compounds,
plus possibly jaconine.

Table II.

Examples of organohalogen compounds as plant elaborates.

Halogen	Compound	Plant	Illustrative reference
Bromine	Br-antibiotics[a]	Microorganisms	5
	Bromophenols	Algae	6
	Bromophenols	Seaweed	7
	Bromoacetic acid	Wine (grapes?)	14
Chlorine	Jaconine	Senecio jacobaea	8
	Several	Microorganisms	4,5
Fluorine	Fluoroacetic acid	2 Dichapetalum spp.	9
Iodine	Iodo-tyrosins	Marine algae	10,11
	Triiodoacetaldehyde	Seaweed	12
	Organic iodine	Wheat	13

[a] Organochlorine-antibiotic producing organisms grown in nutrient
media from which chloride was excluded and bromide was available.

It is suggested that future work to isolate and measure by any
technique organohalogen compounds in plant parts incorporate a step
whereby the organic extract is washed several times also with 10%
nitric acid solution (1) to destroy any lecithin-halide-ion type

complexes that may be present in solution. Depending upon the measuring technique [(e.g., direct potentiometric analysis after combustion (15)], traces of bromine from either organic or inorganic bromides could result in a large error in chloride analysis, as with the cucumber seeds in Table I (cf. footnote e); however, the present data do not consistently demonstrate such a probability.

References

(1) F. A. GUNTHER, J. W. HYLIN, and R. E. SPENGER, J. Agr. Food Chem. (in press).

(2) V. P. GUINN and R. A. SCHMITT, Residue Reviews 5, 148 (1964).

(3) V. P. GUINN and J. C. POTTER, J. Agr. Food Chem. 10, 232 (1962).

(4) J. W. HYLIN, R. E. SPENGER, and F. A. GUNTHER, Residue Reviews (in press).

(5) M. A. PETTY, Bact. Reviews 25, 111 (1961).

(6) P. MASTAGLI and J. AUGIER, Compt. rend. 229, 775 (1949); J. AUGIER and M. E. HENRY, Bull. soc. botan. France 91, 29 (1950); J. AUGIER, Review Gen. Botan. 60, 257 (1953); J. AUGIER and P. MASTAGLI, Compt. rend. 242, 190 (1956).

(7) T. SAITO and Y. ANDO, Nippon Kagaku Zasshi 76, 478 (1955).

(8) J. K. ADDY and R. E. PARKER, J. Chem. Soc. 1963, 915.

(9) R. PETERS and M. SHORTHOUSE, Nature 202 (4927), 21 (1964).

(10) J. ROCHE and Y. YAGI, Compt. rend. soc. biol. 146, 642 (1952).

(11) C. B. COULSON, Chem. & Ind. 1953, 997.

(12) E. MASUDA, J. Pharm. Soc. Japan 55, 625 (1935).

(13)   Z. BÜSZÖRMÉNYI, E. CSEH, and L. GÁSPÁR, Die Naturwissenschaften
       48, 584 (1959); E. CSEH and Z. BÜSZÖRMÉNYI, Plant and Soil 20,
       371 (1964).

(14)   L. CHELLE and G. VITTE, Bull. soc. pharm. Bordeaux 73, 179
       (1935).

(15)   F. A. GUNTHER and R. C. BLINN, Analysis of Insecticides and
       Acaricides, pp. 357-378 (1955), Interscience, New York.

Paper No. 1661, University of California Citrus Research Center and
Agricultural Experiment Station, Riverside, California. Supported
in part by U. S. Public Health Service Grant No. EF-0029 from the
National Institutes of Health.

A Photoisomerisation Product of Dieldrin

by J. Robinson, A. Richardson, B. Bush
"Shell" Research Ltd. Tunstall Laboratory
Sittingbourne, Kent, U.K.

and K. E. Elgar
"Shell" Research Ltd. Woodstock Agricultural Research Centre
Sittingbourne, Kent, U.K.

Roburn (1) reported on the presence of an unknown compound on grass which had been treated with dieldrin and subjected to sunlight. He also showed that the same compound was produced by ultraviolet irradiation of dieldrin on a glass plate.

We have prepared a quantity of this material either by irradiating a solution of dieldrin as described by Bird, et al, (2) or by depositing a thin layer of dieldrin on a sheet of filter paper and irradiating it for 30 minutes with an ultraviolet lamp (2537 Å). The conversion products were extracted with 1:1 v/v hexane/acetone mixture and then applied to preparative thin layer chromatoplates, coated with silica gel H 254. The plates were developed with 1:4 acetone/hexane mixture, and the separated bands were located by u.v. light. The band of the major conversion product was scraped from the plate and extracted from the gel with diethyl ether. Further purification was achieved by chromatography through a silica gel column using hexane/benzene 1:4 as eluant. Recrystallization of the product from ethanol gave a product with a melting point of 188°C.

Using Apiezon L as the stationary phase the retention time of

127

the conversion product (relative to dieldrin) was 4.5 as reported by Roburn; the relative retention times (dieldrin=1) on silicone SE 30, Oronite polybutene 128, QF 1, and neopentyl glycol succinate were 2.6, 4.3, 9, and 20 respectively.

We confirm Roburn's observation that the product is more polar than dieldrin by its behaviour on reversed phase paper chromatography. The R_f values for dieldrin and the photoconversion product on thin layer chromatography were as follows:

TABLE 1

R_f values of dieldrin and photochemical conversion product

Compound	Absorbent	
	Silica Gel (a)	Alumina H (b)
Dieldrin	0.93	0.55
Photoconversion product	0.55	0.46

Developing solvents: (a) 4:1 (v/v) cyclohexane/acetone: (b) 6:1 (v/v) cyclohexane/propanol.

Microanalysis, mass spectrum and molecular weight determinations indicate that the compound is an isomer of dieldrin. Inspection of the infrared spectrum of the conversion product shows that the absorption bands at 6.25 u and 6.81 u, attributable to the dichlorethylene and methylene groups respectively in dieldrin, are absent. The absorption bands attributed to the epoxide group in

128

dieldrin at 8.02, 11.8, and 12.05 u are missing or have been dis-
placed; the appearance of new bands at 8.88 u and 10.57 u may
indicate the presence of an oxygen ring with more than 3 members.
The NMR spectrum of the conversion product confirms the absence of
a methylene group and indicates the presence of a CHCl group.
These observations indicate that a bridge has been formed in the
conversion product between the CH_2 bridge and the Cl-C=C-Cl group
in dieldrin, but the evidence regarding the oxygen function in the
conversion product is ambiguous: the infrared spectrum indicates
the absence of a carbonyl or hydroxyl group and it is tentatively
concluded that an oxygen ring system is still present. This ring
system may be unchanged 1-2 oxide or a larger ring. However, the
presence of a larger ring necessitates the shift of a carbon-carbon
link with the formation of an endo-endo structure. Triphenyl
phosphine has been reported to be a specific reagent for the conversion of 1:2 epoxides to the corresponding ethylene derivatives
but neither dieldrin nor the conversion product reacts with this
reagent. A study of the cracking pattern of dieldrin, endrin
and the conversion product indicates that the pattern of relative
abundance ratios of the breakdown products of the conversion product is similar to that of dieldrin (an endo-exo compound) and
quite different from that of endrin (an endo-endo compound).
This observation indicates that the conversion of the endo-exo
ring fusion in dieldrin under the influence of light to an endo-
endo fusion product is unlikely.

It is tentatively concluded that the structure of the conversion product is:

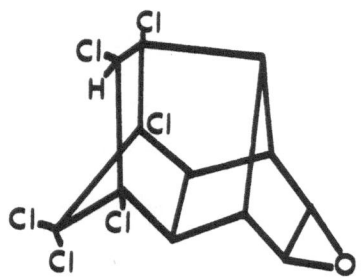

Both dieldrin and the photoconversion product react with anhydrous hydrogen bromide in dioxan to give bromo-hydrins. Both bromo-hydrins are converted back to the original compounds on treatment with alcoholic potash. Once again these reactions are not specific for 1:2 oxide ring systems.

Since the conversion product is formed by the action of sunlight on dieldrin, it was considered essential to obtain evidence of the occurrence of this compound in the environment as a result of current agricultural practice. Direct applications of dieldrin to the edible foliage of plants is uncommon and in this preliminary survey we have concentrated attention upon those crops in which residues of dieldrin have been found in the past. Animal experiments indicate the conversion product has lipophilic properties and specimens or animal fats have also been analysed. In order to assess the overall dietary intake of man, two samples of cooked meals (corresponding to 24 hour samples) have been analysed,

TABLE 2

Concentration of photoconversion product of dieldrin in the environment

Nature of Sample	Number of Specimens	Average concentration of photo conversion product ppm	Lower limit of detection, ppm	Dieldrin content ppm
English mutton fat	2	(a) none detected (b) 0.004	0.001	0.07
Australian mutton fat	1	none detected	0.0001	0.01
Argentine corned beef fat	2	(a) 0.0018 (b) none detected	0.002	0.16 0.015
Crude & re-fined edible oils & fats	8	none detected	0.004 - 0.05	0.05
Whole cooked meals	2	none detected	0.001	0.02
Human fat	10 (pooled)	none detected	0.0005	0.4
Butter	2	none detected	0.001	0.04
Cooked meats	2	none detected	0.001	0.008
Milk	1	none detected	0.001	0.006
Shag eggs	62 (pooled)	none detected	0.0001	2.1
Potatoes	2	none detected	0.005	0.04
Soil	2	none detected	0.01	0.12
Forage beet foliage	1	0.02	0.01	0.09
Forage beet	1	none detected	0.008	0.005

and the pooled body fat from 10 people has also been analysed.

A pooled sample of shag eggs (corresponding to 62 eggs) was included in the specimens as the dieldrin residues in these eggs are higher than those found in the majority of birds eggs.

Apart from the English mutton samples, Argentine beef fat, and the foliage of forage beet, the photoconversion product could not be detected in any of the samples examined. The identity of the component in these samples, estimated as the photoconversion product, requires confirmation.

The results of this preliminary survey of the concentration of the photoconversion product of dieldrin in the human diet indicate that this compound is occurring, if at all, in very small amounts in the environment. The failure to find any of the compound in the bulked sample of human fats, even at a level of less than 0.0005 ppm, also indicates that the concentrations of this compound in the environment are very small. The ratio of the concentration of dieldrin to that of the photoconversion product varies from 16:1 to 1000:1. It is concluded, therefore, that the possible conversion of dieldrin by sunlight to an isomer is not significantly increasing the overall residues arising from the use of aldrin and dieldrin.

1. J. ROBURN, Chem. and Ind., 1963, 1555.

2. C. W. BIRD, R. C. COOKSON, and E. CRUNDWELL, J. Chem. Soc., 1961, 4809.

The Photochemical Isomerization of Dieldrin and Endrin and Effects on Toxicity[1]

by Joseph D. Rosen, Donald J. Sutherland, and Gary R. Lipton
*Department of Agricultural Chemistry and Department of Entomology and
Economic Zoology, Bureau of Conservation and Environmental Science
Rutgers—The State University, New Brunswick, New Jersey*

The commonly-used insecticides dieldrin (I) and
endrin (II) persist in our environment for extended
periods (2). Because these insecticides are exposed to
sunlight during this time, it is important to determine the identity and toxicity of materials being added to our environment by photochemical pathways. Previous studies have shown that both dieldrin and endrin
are readily decomposed by ultraviolet light (3,4). In
addition, it was found that one of the ultraviolet de-
composition products of dieldrin was the same (by paper and gas chromatography) as the material that was
obtained by exposing dieldrin-treated grass to sunlight under natural conditions for several months (4).

These studies extend Roburn's work. A single
photo-conversion product was obtained in yields of 7%
after three weeks and 25% after two months by exposing
dieldrin to sunlight (5). The same compound was obtained in 66% yield by exposing dieldrin to a 2537 A°
germicidal lamp for 48 hours. This material was

Bulletin of Environmental Contamination & Toxicology,
Vol. 1, No. 4, 1966, published by Springer-Verlag New York Inc.

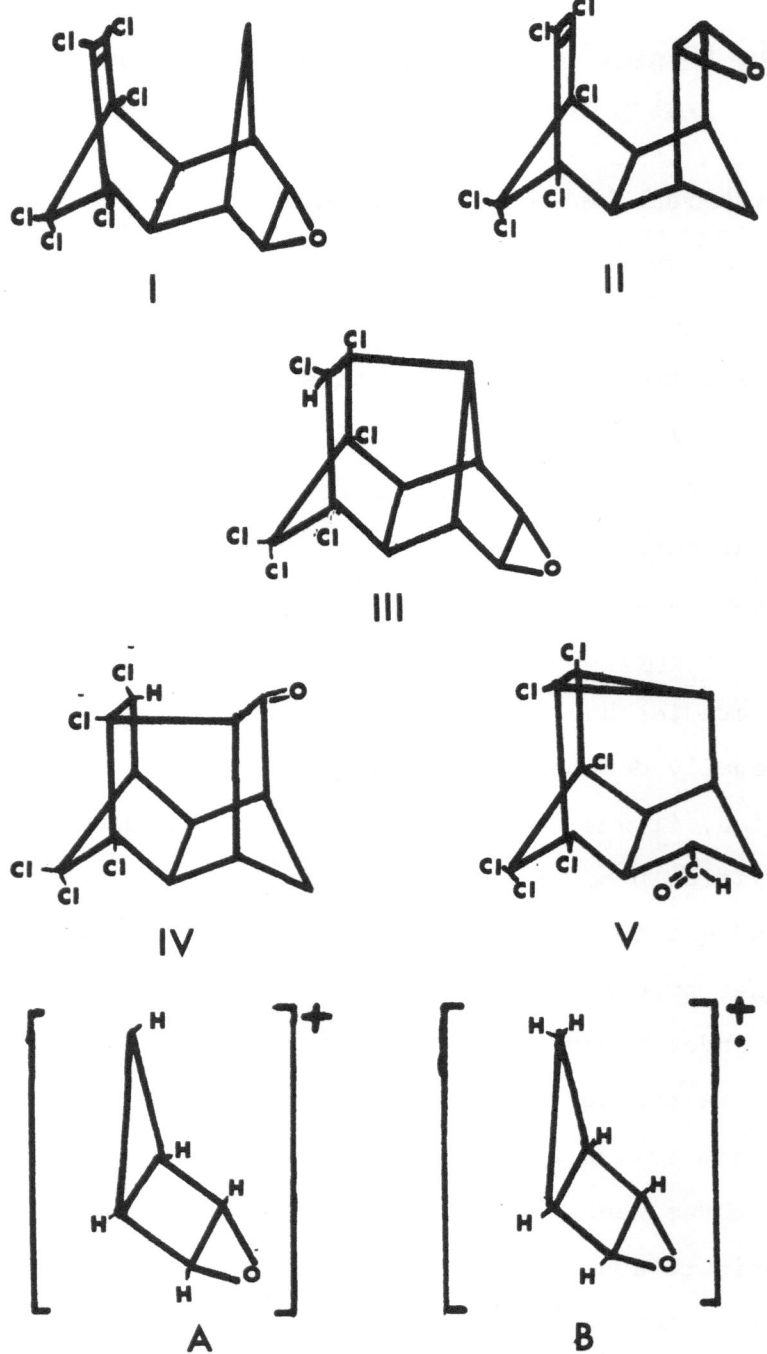

I

II

III

IV

V

A

B

separated from starting material and other minor products by column chromatography on silica gel G, using hexane to elute dieldrin, and hexane-ethyl acetate (7:3) to elute the photo-conversion product. The laboratory and naturally-irradiated products were shown to be identical by comparison of their infrared spectra and by identical GLC and TLC behavior.

On the basis of instrumental analysis and mechanistic interpretation it appears that the photo-conversion product of dieldrin is its hexacyclo isomer, 10-oxa-3,6-exo- 4,5,13,13-hexachlorohexacyclo (6.3.1. $1^{3,6}.1^9.1^{11}.0^{2,7}.0^{5,12}$) tridecane (III). A comparison of the infrared spectrum of III with that of dieldrin showed the disappearance of the chlorinated-olefin peak at 1600 cm^{-1}, the disappearance of methylene absorption at 1470 cm^{-1}, and a small shift in epoxide absorption from 847 to 851 cm^{-1}. The mass spectrum (6) of III exhibited a parent ion at m/e 378 (indicating no change in molecular weight) and a base peak (non-chlorine containing) at m/e 81. This fragment most probably has structure A. In contrast, the mass spectrum of dieldrin exhibited a peak at m/e 82, probably fragment B. There was virtually no peak at m/e 82 in the degradation product. The n.m.r. spectrum (in acetone-d$_6$ with tetramethylsilane as reference) showed ill-defined multiplets at 3.55, 3.44, 3.14, 2.89 and 2.45 δ as well as a singlet at 5.33 δ . We have assigned the latter peak to the migrated hydrogen on the basis of the reported chemical shift at 4.98 δ for the protons of 1,2,3,4,5,6-hexachlorocyclohexane in acetone (7). The ratio of the multiplets to the singlet at 5.33 was nearly 7:1, in accord with the proposed structure of III.

The ultraviolet irradiation of endrin, employing laboratory conditions identical to those used with dieldrin, yield 37% 1,8-exo-9,10,11,11-hexachloropentacyclo $(6.2.1.1^{3.6} . 0^{2,7} 0^{4,10}.)$ dodecan-5-one (IV) and 9% 4,5,6,7,8,8-hexachlorohexahydro-4,7-methano-3,5, 6-methenoindan-1-carboxaldehyde (V). These compounds were isolated by thick layer chromatography on silica gel G by eluting with hexane-ethyl acetate (7:3), and exhibited identical infrared spectra to those of published spectra (8). Endrin was not converted to IV and V by the silica gel. The mechanism of formation of these two carbonyl compounds from endrin by thermal isomerization is thought to involve either a hydride shift or hydrogen abstraction of an epoxy-hydrogen (8). The geometry of the dieldrin molecule precludes involvement of an epoxy-hydrogen, and favors the participation of a methylene hydrogen. Present experiments on the exposure of endrin to sunlight have not been completed but it is likely, on the basis of the preceding discussion, that the carbonyl compounds IV and V will be found.

Compounds I through V were examined for their toxicity to adult house fly, _Musca Domestica_, and larval mosquito, _Aedes aegypti_. The former species included a susceptible laboratory strain (Wilson) and a strain highly resistant to diazinon and less resistant to dieldrin. The laboratory strain of _Aedes_ and its insecticide susceptibility have been reported recently (9). Compounds were applied topically in 1 ul. acetone to 4 day-old adult flies (10); 3 day-old mosquito larvae were exposed to the compounds suspended in water by means of 95% ethanol (9). The LD50 and LD90 in ug/fly and the LC50 and LC90 in p.p.m. for mosquitoes (Table 1) were taken from dosage-mortality regression lines. LD90 values for resistant house flies are not included since the strain is not homogeneous and regression lines are bi-phasic. Compounds IV and V were non-toxic to the house fly and mosquito at concentrations of 0.24 ug/fly and 0.096 p.p.m., respectively.

Based on the results obtained, III is approximately two times more toxic than dieldrin to the house-fly and mosquito. In addition, III was more rapid than dieldrin in producing a toxic response in the house fly. These properties were also noted in the mixture of compounds (29% I, 66% III, obtained by ultraviolet irradiation of dieldrin in the laboratory). The relative toxicities of dieldrin, III and the mixture were approximately the same for both susceptible and resistant house flies, and possibly the site of action and mechanism for resistance for III and dieldrin are similar.

Although some chlorinated hydrocarbons may owe their activity to the formation of a charge-transfer complex (11), it seems that dieldrin would have less potential than DDT for such formation. Although III is more polar on the basis of solubility, by virtue of its proposed structure, it would seem to have less potential for forming charge-transfer complexes than dieldrin. Therefore, the increased toxicity and speed of action of III may be due not to a greater potential for complex formation, but to a more rapid and thorough penetration to the site of action and/or to a more precise fitting into hypothetical intermolecular lattices (12). Dieldrin has a topical:injection LC50 ratio (house fly) for 24 hours of 1.7 (13) indicative that dieldrin is relatively efficient in reaching the site of action. Studies are in progress to determine if ·III is actually more efficient than dieldrin.

Information on the mammalian toxicity of III and methods for its residual detection are forthcoming.

TABLE 1

$LD_{50}(LD_{90})$ ug/fly to Musca domestica and $LC_{50}(LC_{90})$ p.p.m. to _Aedes_ _aegypti_ of compounds.

	Hours	Dieldrin (I)	Dieldrin Isomer (III)	Mixture
Susceptible house fly	1	>.24	<.173	.09 (.2)
	2	>.24	.043 (.086)	.05 (.1)
	6	.035 (.07)	.01 (.019)	.018 (.06)
	26.5	.015 (.029)	.006 (.013)	.012 (.035)
Resistant house fly	6		.034	.058
	26.5	.032	.017	.025
Susceptible mosquito	24	.0058 (.0084)	.0029 (.0046)	

REFERENCES

1. Paper for the Journal Series, New Jersey Agricultural Experiment Station, New Brunswick, New Jersey. This reserach was supported by U.S.P.H.S. Research Grant #ES-00016 from Bureau of State Services.

2. Our gratitude to Shell Development Corp. for their generous gift of compounds I, II, IV, and V.

3. L. C. MITCHELL, JAOAC, 44, 643 (1961).

4. J. ROBURN, Chem. Ind., 38, 1555 (1963).

5. We are grateful to W. B. DEICHMANN, University of Miami, for the natural exposure of dieldrin.

6. Performed by Morgan-Schaeffer Corp., Montreal, Can.

7. R. K. HARRIS and N. SHEPPARD, Mol. Phys., 7 (6), 595 (1963-4).

8. D. D. PHILLIPS, G. E. POLLARD, and S. B. SOLOWAY, J. Agr. and Food Chem., 10, 217 (1962).

9. D. J. SUTHERLAND, N. J. Mosq. Exterm. Assoc., 51, 107 (1964).

10. A. J. FORGASH and E. J. HANSENS, J. Econ. Ent., 55, 679 (1962).

11. R. D. O'BRIEN and F. MATSUMURA, Science 146, 657 (1964).

12. L. J. MULLINS, Science 122, 118 (1955).

13. A. J. FORGASH, unpublished information.

A Sodium Flame Detector of Increased Stability for Phosphorus-containing Pesticides[2]

by DAVID R. COAHRAN
Department of Agricultural Chemistry
Washington State University, Pullman, Washington

One of the chief difficulties in analytical gas chromatography is the identification of the resolved components of a mixture. The use of selective detectors sensitive to only a limited class of compounds can simplify this problem considerably.

Giuffrida and Karman have reported the development of such a detector (2,3,4), sensitive chiefly to compounds containing phosphorus and chlorine. A flame ionization detector is used, modified by placing a small amount of an alkali metal salt, usually a sodium salt, in such a position that it is heated by the flame. In some versions of this detector, one electrode is coated with the fused salt. In others, a probe containing the salt is placed in or near the flame.

---

[2]Scientific Paper 2657, Washington Agricultural Experiment Stations. Work was conducted under Project 1793.

141

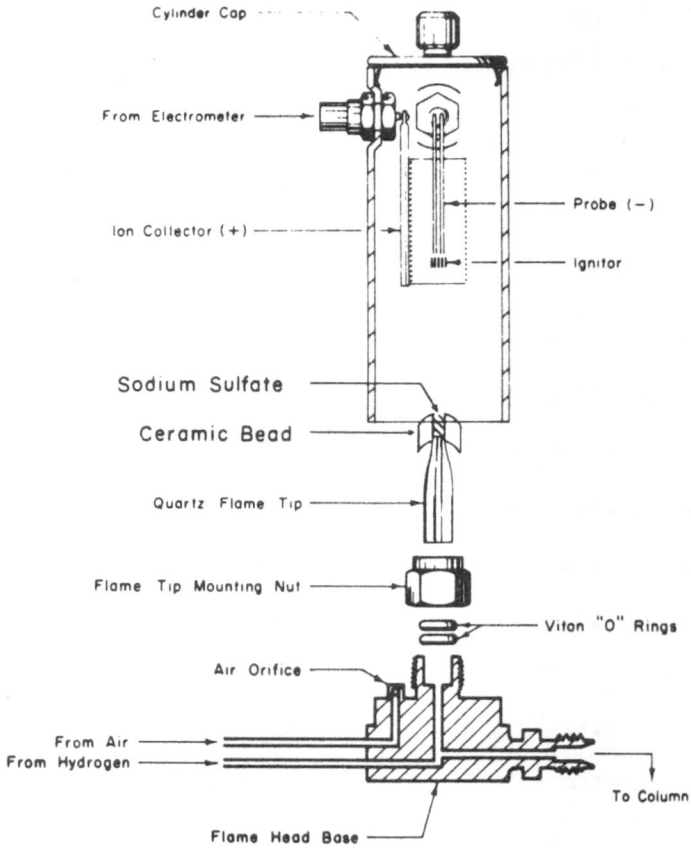

Figure 1.   Construction of the sodium flame detector.

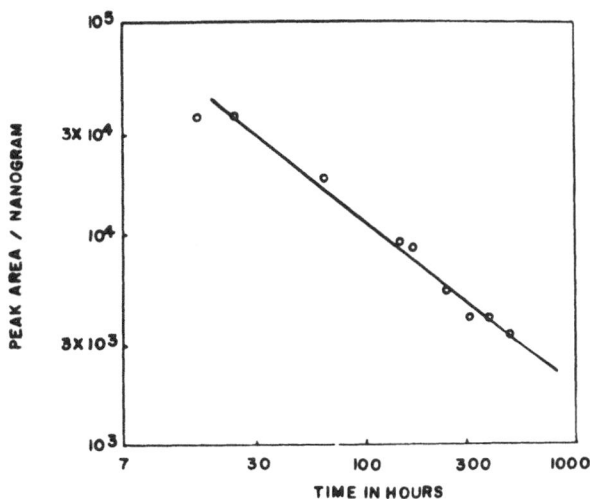

Figure 2.   Detector response vs time.

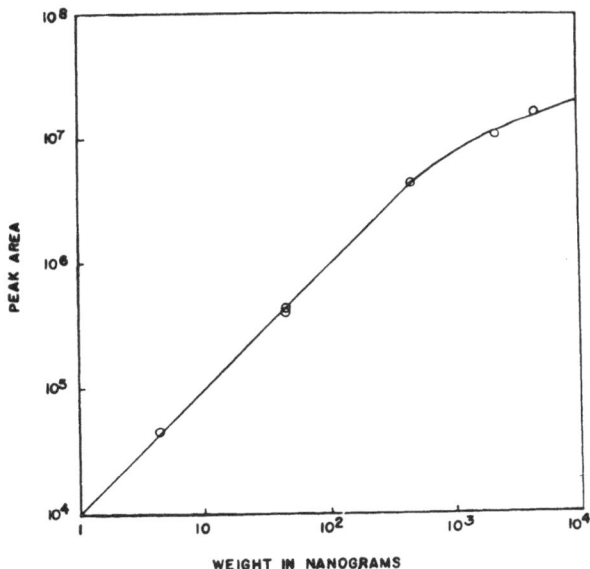

Figure 3.   Detector response vs sample weight.

Several detector designs similar to those described by Karman and Giuffrida were tried with disappointing results. The thin salt coating was rapidly consumed, causing a large change in sensitivity within a few hours. Eventually we developed a much more stable s stem, employing a modified Wilkens hydrogen flame detector. A ceramic tube[3] is placed around the jet (Figure 1). The cup so formed is filled with granular anhydrous sodium sulfate. The mixture of hydrogen, carrier gas and sample flows up through the salt bed and burns at its surface.

The response of this detector drops much more slowly than that of any other design yet tried by us. The detector can be used several weeks between fillings with salt granules (Figure 2). The rate of variation in response is comparable to that of an electron capture detector but is systematic instead of random. The limit of detection falls off less rapidly than the response, since baseline noise, as well as response, decreases with time. Figure 3 is a typical calibration curve. It can be seen that there is a thousand fold range of useable concentrations.

This detector has been compared with an unmodified Wilkens hydrogen flame detector for a representative group of compounds.

---

[3] A ball and socket bead, #P189, inside diameter .29 cm. length .66 cm., manufactured by Saxonburg Ceramics, Saxonburg, Pennsylvania.

The results are shown in Table 1. A Wilkens Instrument and
Research, Inc. Model A-600-C chromatograph with a 5' x 1/8"
column of 5% Dow 11 on 60/80 mesh Chromosorb W was used. The
carrier was Matheson prepurified nitrogen at 30 ml/minute.
Hydrogen flow to the flame was 49 ml/minute. The column temperature was 188-190°C. All peaks were nearly symmetrical with
very slight tailing.

TABLE 1

Comparative Response of Hydrogen and Sodium Flame Detectors

Weights and areas are the averages of several samples.
Peak area is the product of peak height in recorder divisions at
maximum electrometer sensitivity and width in seconds at half
height. With the electrometer used, one division at maximum
sensitivity represents 4.1×10^{-14} ampere, making a unit of area
4.1×10^{-14} coulomb.

| Compound | Flame | | Sodium Flame | | Response |
	wt, ng	area/ng	wt, ng	area/ng	Ratio
anthracene	74.4	26	21660	13	0.49
hexadecane	196	32	3737	15	0.48
ethyl stearate	1050	27	153000	5.2	0.19
Morestan	504	12	4490	71	5.8
aldrin	410	17	4044	1000	59
PCNB	345	9.2	3318	960	100
malathion	659	2.3	463	8600	3800
parathion	448	13	43	37,000	2800
Di-Syston	498	15	52	52,000	3400
Diazinon	280	24	26.7	75,000	3100
Phosphine			0.38	150,000	

As can be seen from Table 1, the ratios of the response of the sodium flame detector to those of the hydrogen flame detector fall into three groups. Compounds not containing phosphorus or chlorine have ratios near 1. The lower response per nanogram of the sodium flame detector to some of these materials may be due to the much greater sample size required. Compounds containing approximately 60% chlorine, such as aldrin and pentachloronitro-benzene (PCNB), give response ratios of about 100. Those containing about 10% phosphorus give ratios near 3000. Giuffrida and Karman have quoted various response ratios ranging up to 20 for chlorine and 600 for phosphorus compounds. It is evident that the cup type sodium flame detector has enhanced discrimination against phosphorus- and chlorine-free compounds, as well as improved stability.

To test the sensitivity of this detector to reduced phosphorus, an attempt was made to obtain the response ratio for phosphine. Although the sodium flame detector works well with phosphine, the response of the hydrogen flame detector is very erratic. It is suspected that the intrinsic sensitivity of a hydrogen flame detector to phosphine is very low, and that the observed response is due largely to contamination of the detector by traces of sodium salts, which are extremely difficult to remove.

Although both the baseline current and the noise level are much higher for the sodium flame detector than for the unaltered hydrogen flame detector, for organophosphates the smallest detectable quantity (that is, the limit of detection), is lower. The increase in noise is more than compensated for by the increase in Dimbat-Porter-Stross "sensitivity" (1). The increased baseline current may require a higher "bucking" current which can be obtained by increasing the voltage of the bucking battery or, in the case of a Wilkens electrometer, by setting the "range" switch to an "EC" position. (1) Baseline current, noise, and "sensitivity" all decrease slowly with time since replenishing the salt supply.

A typical noise level for the sodium flame detector is 1000 divisions at maximum electrometer sensitivity. The hydrogen flame detector sometimes gives a noise level as low as one division, but 20 to 50 divisions is not rare. Since the enhancement ratio for the phosphorus compounds tested is about 3000, the sodium flame detector has a limit of detection at least three times better than the hydrogen flame detector. Injections of four nanograms of Parathion or Malathion or of one nanogram of Diazinon or Di-Syston give clearly visible peaks. As can be seen from figure 3, the response curve is nearly linear up to several thousand nanograms.

We have used this detector for over a year in the analysis
of soil and plant tissue samples for organophosphorus pesticides.
Very few phosphorus-free compounds appear on chromatograms made
with it and these can be easily identified as such by the use of
a hydrogen flame detector. Its selectivity has made gas chromatographic analysis possible in some cases where plant materials
previously interfered. In addition it often allows the use of
simplified cleanup procedures, saving time and materials and
giving improved recoveries.

References

1. M. Dimbat, P. E. Porter and F. H. Stross, Anal. Chem. 28, 290,
   (1956).

2. L. Giuffrida, Journal A.O.A.C. 47, 293, (1964).

3. A. Karman and L. Giuffrida, Nature 201, 1204, (1964).

4. A. Karman, Anal. Chem. 36, 1416, (1964).

The Characteristics and Operation Parameters of a Thermionic Emission Detector, Selective and Sensitive to Phosphorus

by HERMAN BECKMAN and WILLIAM O. GAUER
Agricultural Toxicology and Residue Research Laboratory
University of California, Davis, California

The thermionic emission detector, a newcomer in the line of gas liquid chromatograph detectors, is rapidly proving to be a valuable tool for the specific detection of organophosphorus compounds.   The greatest impact is presently in the pesticide field.   At best, the detector is still plagued with some operational difficulties and detection inadequacies, leaving considerable room for its improvement.   In addition, a complete theoretical description of the flame reaction mechanism is also still lacking.   It is hoped that most, if not all of these difficulties can be overcome in the future as new research ideas are successfully applied.   This paper is written to help the analyst, unfamiliar with the operations of a thermionic detector to avoid possible difficulties in its operation, as well as to stimulate new ideas to improve existing systems.   For these reasons a literature review will be included.

1.   Review

A hydrogen flame ionization detector for gas chromatography was first reported in 1958 by Harley, Nel, and Pretorius

149

(1), and McWilliam and Dewar (2), the former describing a
single flame system and the latter both single and dual flame
systems. Since that time these detectors have been extensively used and are well known for their relatively high
sensitivity to compounds capable of thermally generating
carbion ions in the flame, and their lack of response to most
other ionic species.

In March of 1964 Karmen and Giuffrida (3) reported a
procedure for selectively increasing the responsiveness of
a flame detector to compounds which contain chlorine, bromine, iodine, or phosphorus. A circular electrode coated
with sodium hydroxide was suspended 5 mm above the flame
and ninety volts were impressed between it and the body of
the electrode. The resulting arrangement gave rise to an
enhancement of the response to compounds containing halogens and phosphorus. In April of 1964, Giuffrida (4)
described in detail the construction and optimal operation
parameters of a similar specific detector, referred to as a
Sodium Thermionic Detector. A coating of sodium sulfate
was fused onto the electrode and a 300 volt battery was used
to provide the detector voltage. The response on the thermionic detector to an organic halide was found to be greater
than that of the conventional flame detector but considerably
less than that produced by an organic phosphorus compound
on the same thermionic detector. Some pesticide detection
data was also included. In July of 1964, Karmen (5) reported a study of a single and a double flame thermionic
detector. The single flame detector was constructed and
operated similarly to those described previously. The
double flame detector had two flame jets, one above the
other with an alkali metal hydroxide or salt coated wire
mesh screen between them. A particular advantage of the
two-flame system noted was the complete insensitivity of the
detector to non-phosphorus or non-halide containing organic

compounds, once they had passed through the first flame. Near
the end of 1964, Giuffrida and Ives (6) reported the use of a
thermionic detector in an investigation of recovery procedures
for several organophosphorus pesticide residues on various
crops. The thermionic detector was used in conjunction with a
flame ionization detector, the former providing quantitative
pesticide recovery data and the latter providing cleanup efficiency
data.

In June of 1965, three studies of interest involving the
thermionic detector were reported. Coahran (7) converted a
commercially available flame ionization detector to a thermionic detector by placing a ceramic tube over the flame tip and
filling the "cup" thus formed with granular sodium sulfate. The
emerging gases from the flame tip passed through the salt bed
and burned on its surface. Greater long term stability was
claimed for this system. Schmit, Wyme, and Peters (8) reported achieving good results by coating a copper electrode with a
mixture of cuprous nitrate, boric acid, and acid silver solder
flux. The electrode coil with its glassy coating was placed
over the flame tip, and the remaining wire was attached to the
lower end of the metal flame tip itself. The claims of the
report were the development of a low noise, long life coating
while preserving the sensitivity and specificity of the original
detector. At this time our laboratory (9) reported the conversion of an existing flame ionization detector to a thermionic
detector similar to Giuffrida's design (4). Contrary to some
existing pessimism over the short-term stability of this detector
design, it was found that with critical electrode placement and
careful gas control, a system of comparable sensitivity and
relatively long term stability could be devised.

In October of 1965, Karmen (10) reported a study involving the relative sensitivities of several halide and phosphorus

containing organic compounds in the presence of interfering
organic materials, in the double flame thermionic detector.
Data on the comparison of several salt coatings were given with
cesium chloride being found to be the most highly sensitive. In
early 1966, Giuffrida, Ives and Bostwick (11) reported on an
investigation of special ionization detectors which included operational parameters of their thermionic detectors. A study was
made on several salt coatings, the results indicating that salts
of potassium were of the greatest sensitivity. Thermionic and
electron capture dual analysis data for several pesticides are
also given.

Theoretical treatment of the flame mechanism of this
unique detector is still lacking in the literature, to date. Padley,
Page and Sugden (12), in their theoretical description of the
mechanism of steady state ionization in flames, describe the
special effect of halogens on the ionization in alkali-laden hydrogen flames. No similar treatment has been noted for phosphorus.

2. Experimental

The instrument used in this work is the Wilkens instrument
and Research, Inc., Aerograph Hy-Fi Model 600 with flame
ionization detector (FID).

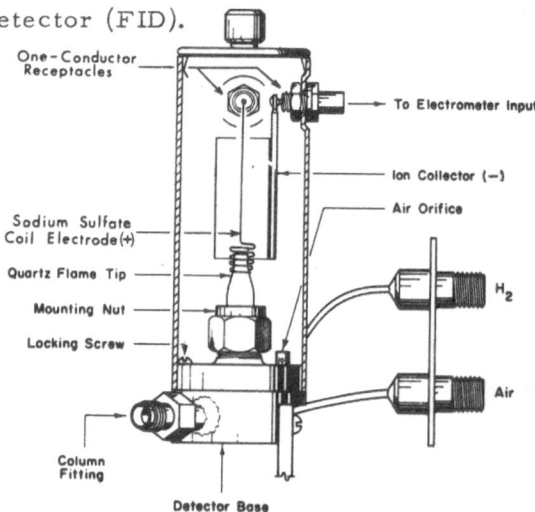

Figure 1.Sodium Thermionic Detector Head

152

The FID detector cell is converted to a sodium thermionic detector (STD) and is shown in Figure 1. Remove the two-conductor ignitor assembly from the FID cylinder and replace it with a one-conductor receptacle the same as that used to accommodate the ion collector. Make a five-turn coil out of nichrome or platinum wire, turning it on a 10-32 screw. Attach the wire to the newly-installed receptacle and center the coil 1-2 mm above the flame tip. Coat the coil with sodium sulfate[1] by successive applications of the saturated solution and warming over a low flame. It is convenient to remove the cylinder, leave the electrodes attached, and apply the salt solution with an eyedropper.

A standard one-conductor electron capture cell cable that contains the 470K resistor inside the cable plug connects the STD cell collector electrode to the 300v ionization output terminal of the instrument. This cable allows the cell to be operated at 90v which was found to be within the optimum voltage range for this system. The Wilkins Model 630 E. C. voltage control or equivalent is recommended where more exact voltage requirements are needed. Recent experiments have indicated that substituting a 90v "B" radio battery (+ to chassis ground) for the power supply will reduce the background noise somewhat at high sensitivity settings. A standard one-conductor FID cable connects the salt-coated electrode to the electrometer input terminal.

A large portion of the detector signal "noise" can be attributed to fluctuations in the H_2 gas flow to the flame. Therefore it is essential that some adequate means of gas regulation

---

[1]We have recently found sodium or potassium borate to have better coating characteristics than the sulfate, with no decrease in sensitivity.

be included if high detector sensitivities are desired. A good, low pressure, precision regulator in series with a two-stage tank regulator functions well. It is also essential to include an indicating controller of the floating ball type or equivalent to insure reproducibility of flow rates. The Wilkins Model 650 hydrogen generator was found unsuitable for this application. The N_2 carrier gas flow seems to be sufficiently uniform and requires no special attention. Air flow was maintained as delivered from an aquarium pump and molecular sieve trap.

Operation of this detector is similar to a standard FID. Since the ignitor assembly was removed at the time of conversion, the H_2 is ignited with a flint striker. Carrier and H_2 flow rates should be 40 mls/min. and 50 mls./min. respectively. Attenuator settings are relatively high. A total signal attenuation of 32,000 x will produce a one-half scale recorder response to one microgram of parathion. Detection limits are one nanogram absolute minimum and approximately 10 nanograms being a workable limit.

3. Results and Discussion

The responses of the thermionic detector with different sodium salt electrodes at optimum hydrogen flow are shown in Figure 2. The numbers on the top of the bars refer to the relative hydrogen flow rates needed for maximum response. Although sodium chloride gave the highest response, it was found to be quite erratic at high sensitivities. The importance of hydrogen flow stability is self-evident in Figure 3. A small change in flow rate markedly affects the recorder response. In addition, the response of the phosphorus compound and that of the solvent are affected quite differently by the change in hydrogen flow, demonstrating the difference in the effect of the two substances on the alkali-laden flame. The cell voltage affects the response from the detector to 10 nanograms parathion as shown in Figure 4. Although higher sensitivities

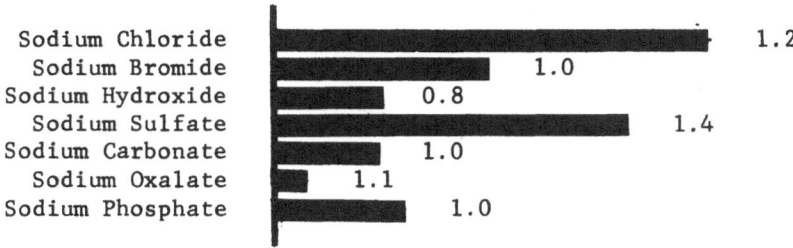

Figure 2.   STD Electrode Salt Coating Response

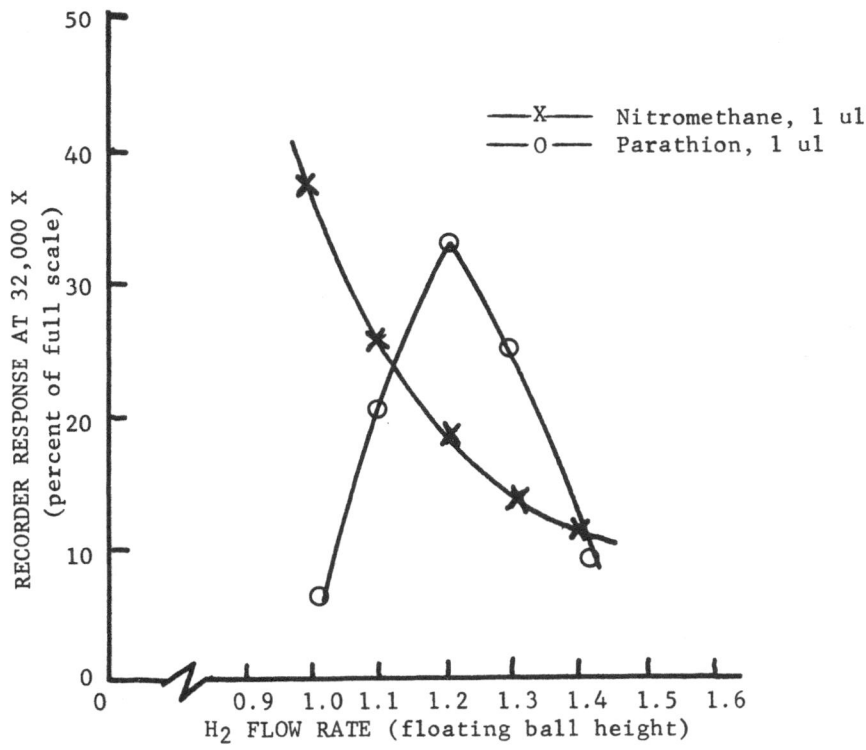

Figure 3. STD H$_2$ Flow Versus Response

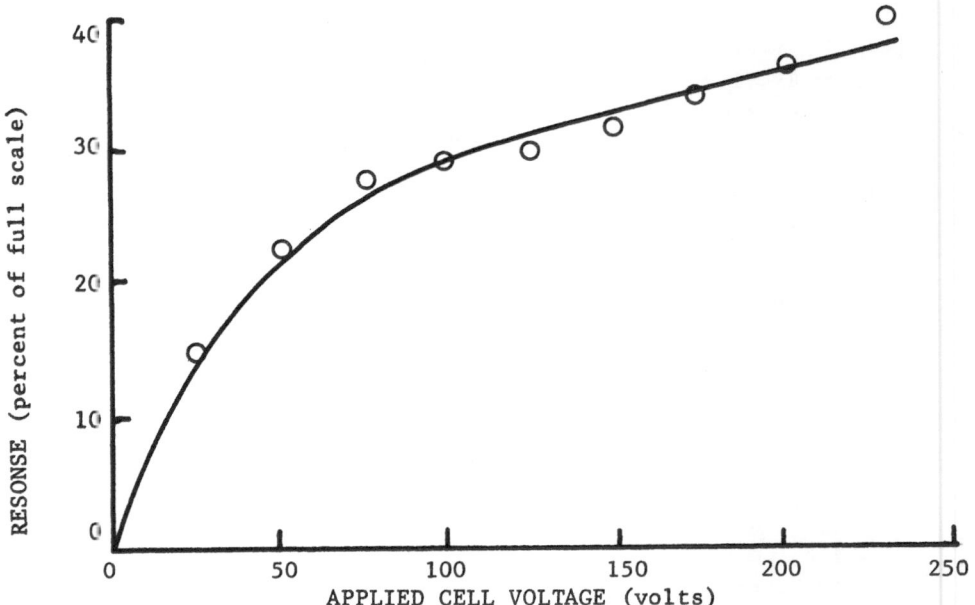

Figure 4. STD Response Versus Applied Voltage

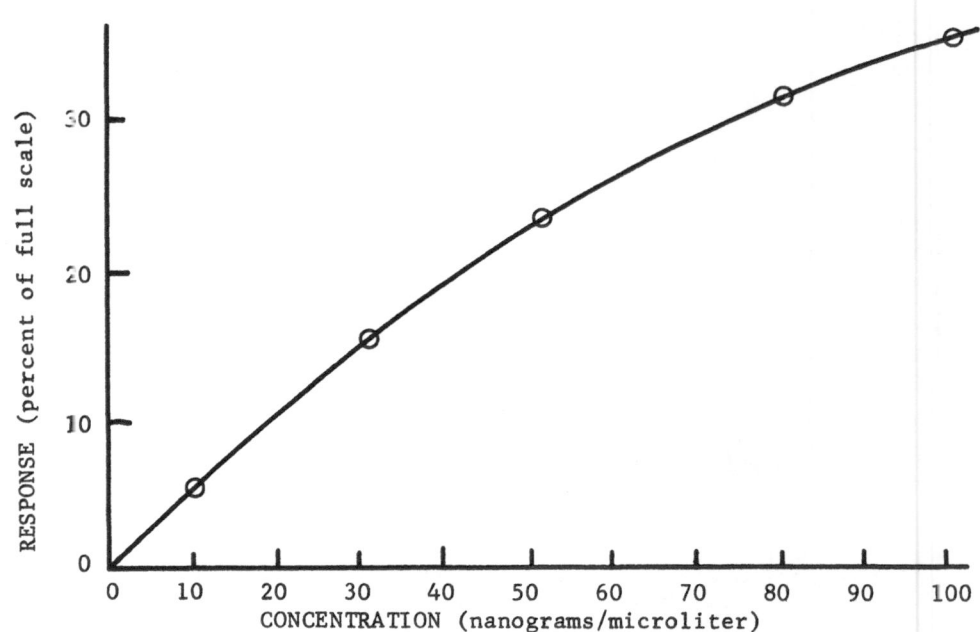

Figure 5. STD Response Versus Concentration

are achieved at higher voltages, the signal/noise ratio appears
to improve in the direction of lower voltage. A concentration
versus recorder response curve for parathion is shown in Figure
5. The curve is non-linear, indicating the need to plot a standard curve in quantitative analysis. A comparison of responses
of several solvents on both the thermionic and flame ionization
detector are shown in Figures 6 and 7. One microliter of each
solvent was injected under the same column conditions. A one
microgram parathion sample was also included. As indicated
by this study the selection of a solvent for use with a thermionic
detector differs considerably from that for use with flame
ionization. Normally, where hexanes, benzene and similar
solvents are not used with flame ionization detection, they work
quite well with thermionic detection. The reverse is true for
the chlorinated solvents.

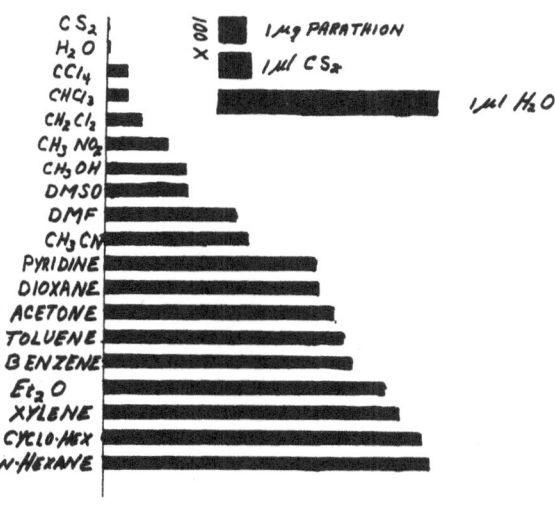

Figure 6. FID Solvent Response

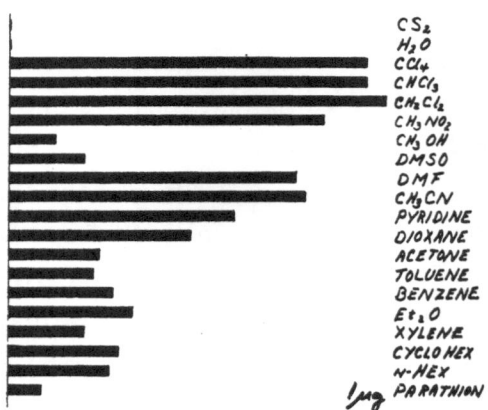

Figure 7. STD Solvent Response

References

1. J. Harley, W. Nel and V. Pretorius, Nature **181**, 177 (1958).

2. I. G. McWilliam and R. A. Dewar, Ibid. , 760 (1958).

3. A. Karmen and L. Giuffrida, Ibid. , **201**, 1204 (1964).

4. L. Giuffrida, Journal A. O. A. C. **47**, 293 (1964).

5. A. Karmen, Anal. Chem. **36**, 1416 (1964).

6. L. Giuffrida and N. F. Ives, Journal A. O. A. C. **47**, 1112 (1964).

7. D. R. Coahran, paper presented at ACS regional meeting, Corvallis, Ore. , June 1965.

8. J. A. Schmit, R. B. Wynne and V. J. Peters, Biomedical GC Notes No. 5, F and M.Scientific Corp. , Avondale, Pa. , June 1965.

9. H. Beckman and W. O. Gauer, paper presented at Wilkins Instrument and Research, Inc. symposium, Walnut Creek, Ga. , June 1965.

10. A. Karmen, J. Gas Chromatog. **3**, 336 (1965).

11. L. Giuffrida, N. F. Ives and D. C. Bostwick, Journal A. O. A. C. **49**, 8 (1966).

12. P. J. Padley, F. M. Page and T. M. Sugden, Trans. Faraday Soc. **57**, 1552 (1961).

Phosphorus Detector for Pesticide Analysis

by C. Harold Hartmann

Varian Aerograph, Walnut Creek, California

A new detector for the gas chromatograph has been developed (4).
It has the unique characteristic of being extremely sensitive to phosphorus
containing compounds while simultaneously being insensitive to all other
organic materials. These characteristics ideally suit the requirements
for the analysis of phosphorus containing pesticides. This paper will
describe the detector geometry, the operation and the performance
characteristics including specificity, sensitivity, linearity and durability.
Also included will be a demonstration of temperature program capability
with the organo phosphorus pesticides.

Description of Geometry

The Aerograph Phosphorus Detector is basically very simple in
design. A cross-sectional view of the detector assembly is shown in
Figure 1. It consists of a standard hydrogen flame detector plus the
addition of a small alkali metal salt pellet. Also added is a base extension
for mounting purposes. The salt pellet consists of one gram of cesium
bromide plus a suitable filler pressed under high pressure to form a
rugged ceramic-like pellet. The base extension serves the dual function
of repositioning the ignitor coil, when changing from the standard FID

159

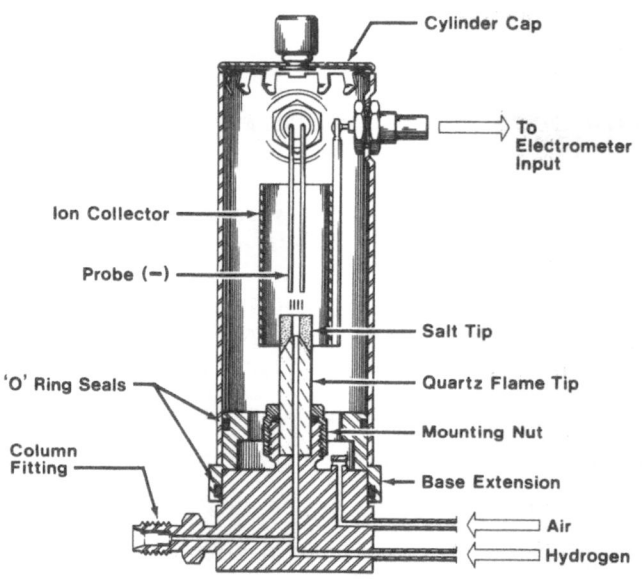

Figure 1 - Aerograph Phosphorus Detector

to Phosphorus Detector operation, and of sealing the base of the detector chamber from extraneous back-diffusion of air into the detector.

Operation

The operation of the Phosphorus Detector is much the same as the standard Flame Ionization Detector: a small hydrogen flame burns on top of the burner tip and ion products from combustion are collected. The similarity ends here as the standard FID uses a quartz burner tip and the Phosphorus Detector uses a cesium bromide burner tip. The background current of the FID is usually about 10^{-11} amps and the Phosphorus Detector is about 3×10^{-9} amps. As a matter of operating procedure the H_2 flow is set so that this high background current is generated. This background current is generated by the ionization of the cesium bromide by the hydrogen flame. When a phosphorus-containing compound passes through the flame a further increase in collected ionization current is measured. The mechanism for this increased ionization is not known.

Flow Sensitivity - The detector has the undesirable characteristic of flow sensitivity. The air flow rate, required to support combustion and purge the detector chamber, requires control to ± 0.1 ml/min. of the 170 ml/min. used for normal operation. The hydrogen flow needs to be controlled to ± 0.01 ml/min. of the 14 ml/min. used for optimum performance. The absolute flow rate of H_2 or Air is not so important, but it must be carefully controlled. The flow control required for analytical purposes can be obtained with a high quality flow controller.

Performance Characteristics

Specificity - One of the unique characteristics of the Phosphorus Detector is that the detector approaches a classic ideal for a gas chromatographic detector in that it can be "tuned" to a particular class of compounds (4). Figure 2 demonstrates, qualitatively, the response of the detector to three different classes of compounds: phosphorus-containing (A), chlorine-containing (B), and normal hydrocarbons (C). As can be seen, the response to chlorinated compounds and normal hydrocarbons can be positive, zero, or negative depending on the particular hydrogen flow used. The response for phosphorus-containing compounds shows a sharp decline above 16 ml/min. because of the increased noise level associated with these flow rates. Although the peak height response increases, the noise level increases more rapidly to give a net decrease in minimum detectability.

161

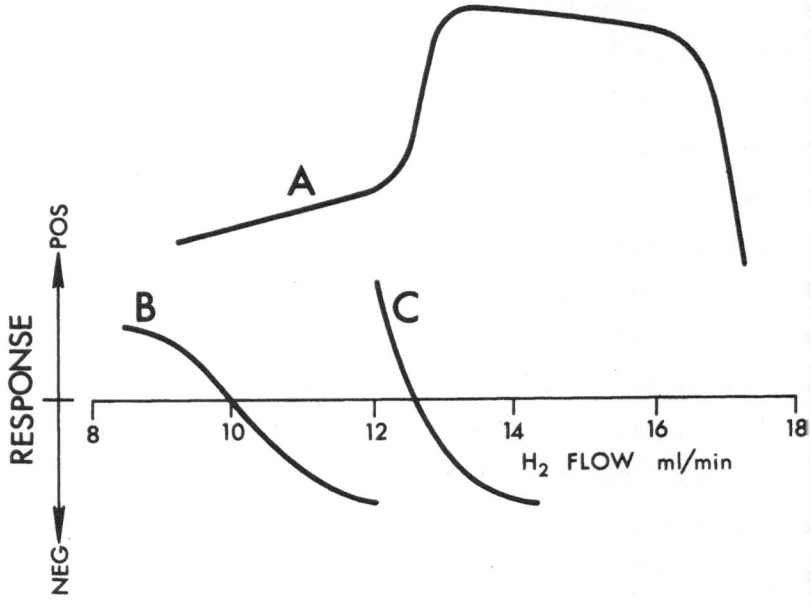

A.   Phosphorus Compounds
B.   Chlorinated Compounds
C.   Normal Hydrocarbon Compounds

Figure 2 – Selectivity of Phosphorus Detector

Sensitivity - Besides being ideally selective, the Phosphorus
Detector more than adequately meets the sensitivity requirements
normally imposed by pesticide residue analysis. Table I lists nine
common organo phosphorus pesticides along with two figures for
sensitivity to the Phosphorus Detector. The first column is the
commonly used sensitivity expression **gm/sec.** which is determined
by dividing the minimum detectable quantity in grams by the width of
the peak in seconds. This sensitivity expression is an attempt to
discount the fact that most compounds have different retention times.
Since it is impractical to change conditions to elute every pesticide in
the same time, the second column states the minimum detectability

in picograms using typical isothermal operating conditions. If these amounts were contained in one microliter of injected extract, they convert directly to ppb, i.e., 3 pg Thimet in one microliter equals 3 ppb minimum detectability.

TABLE I

Phosphorus Detector Sensitivity

Compound	Minimum Detectability gm/sec.	picograms	Retention Time Time (min.)
1 Thimet	$2x10^{-13}$	3	1.9
2 Di-Syston	$2x10^{-13}$	5	2.9
3 Methyl Parathion	$3x10^{-13}$	8	3.8
4 Parathion	$3x10^{-13}$	12	5.5
5 Malathion	$3x10^{-13}$	12	5.5
6 Ethion	$8x10^{-13}$	70	15.2
7 Trithion	$5x10^{-13}$	85	17.1
8 EPN	$4x10^{-13}$	100	26.7
9 Co-Ral	$7x10^{-13}$	500	69.8

Linearity - The typical range of linearity of the Phosphorus Detector is about 1000 fold. This refers to the range of concentration measured from minimum detectability to the highest concentration still yielding a proportional response to sample input. For example, the linear range of the pesticide parathion is from 10 picograms to 10 nanograms. This 10 nanogram upper limit of linearity does not mean that the detector ceases to function at higher concentrations, but that higher concentrations will yield a non-proportional response to increased sample load.

Durability - The most severe weakness of previously reported detector designs (1) (3) has been the poor reliability of the detector performance. The coated wire and screen required several hours to equilibrate and suffered continuous sensitivity deterioration throughout its short lifetime. The Aerograph Phosphorus Detector, however, equilibrates in a matter of a few minutes, has an expected lifetime of thousands of hours, and maintains constant sensitivity throughout an 8-hour work day.

Experimental

The right-hand column of Table I would suggest that temperature programming might be used to good advantage to reduce the time of analysis and to improve the minimum detectability of Ethion, Trithion, EPN, and Co-Ral. Curve C of Figure 2 would suggest that temperature programming might be suggestful since the detector is insensitive to most materials normally associated with column bleed and temperature programming problems.

Apparatus - Figure 3 shows the front view of the gas chromatograph and flow control module used. The lower left module is the linear temperature programmer. The right-hand module is the dual differential electrometer (only channel "A" was used). Not shown at the back of the instrument is the separate detector oven with the protective detector insulator. Although this instrument can be used with dual column detectors, it has been used here with single column and single detector.

Conditions - The column used was a 5' x 1/8" Pyrex glass filled with 5% Dow-200 on Aeropak 30, 70/80 mesh. The temperature was programmed from 190°C to 250°C. The detector and injector temperatures were held isothermal at 200°C. The flow rates were 20 ml/min. nitrogen through the column, 170 ml/min. air to support combustion, and 14 ml/min. hydrogen. The pesticides were Thimet, Di-Syston, Methyl Parathion, Parathion, Trithion, Ethion, EPN, and Co-Ral. The

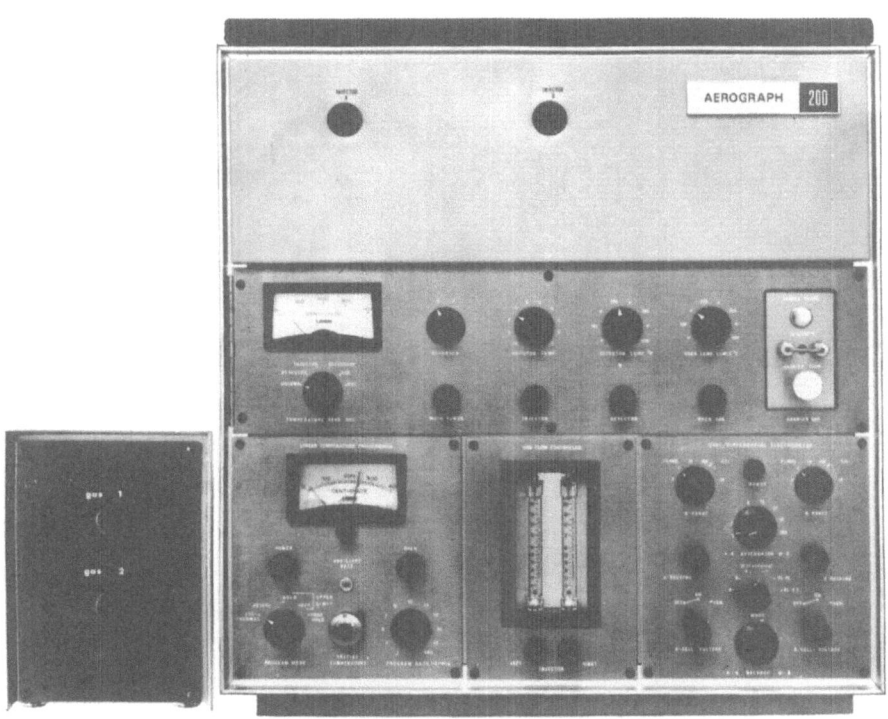

Figure 3 - Modular Analytic GC Ionization Detector - Dual Column

amount injected was 0.25 microliters of 2 ppm solution each, except
4 ppm EPN and 12 ppm Co-Ral. The attenuation is indicated on
Figure 4 (1x = 4 x 10^{-10} amps full scale).

Results and Discussion

Figure 4 shows a typical temperature programmed pesticide analysis
of the eight phosphorus-containing compounds. For this analysis to be
done isothermally and achieve the same separation of early peaks would

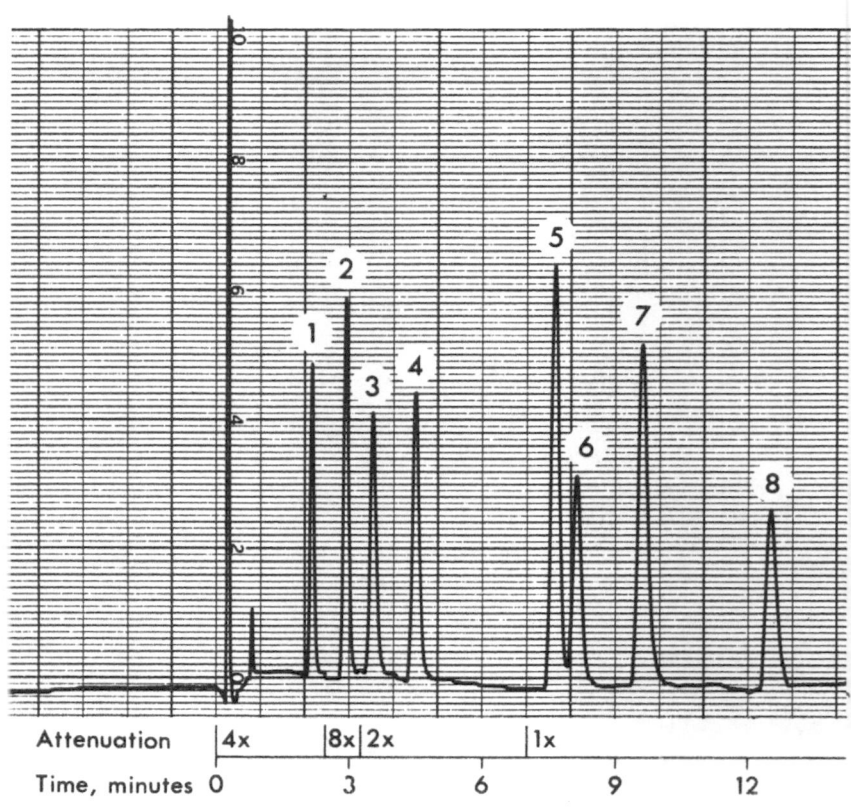

Figure 4 - Temperature Program of Phosphorus Pesticide

No.	Compound	Nanogram Amount	Conditions
1	Thimet	0.5	Column: 5'x1/8", 5%
2	Di-Syston	0.5	Dow-200
3	Methyl Parathion	0.5	Column Temp: 190°C
4	Parathion	0.5	to 250°C
5	Trithion	0.5	Injector Temp: 200°C
6	Ethion	0.5	Detector Temp: 200°C
7	EPN	1.0	N_2 Flow: 20 ml/min.
8	Co-Ral	3.0	H_2 Flow: 13 ml/min.
			Air Flow: 170 ml/min.

require 71 minutes. The programmed analysis time was 13 minutes or 5-1/2 times faster. The accuracy of the temperature program is shown in Table II. The eight peaks are listed and their respective retention times given for 30 consecutive analyses. The relative standard deviation is calculated for each of the peak retention times and, as can be seen, represents high reliability for identification purposes. Previously the electron capture detector was the only detector capable of picogram detection of pesticides, but it could not be programmed because of its sensitivity to column bleed which changes with temperature. Although the Phosphorus Detector is only suitable for the analysis of phosphorus-containing compounds, at least this group is capable of this very useful gas chromatographic technique.

TABLE II

Peak No.	Compound	Average Retention Time (min.)	Relative Stnd. Dev.
1	Thimet	1.930	3.6%
2	Di-Syston	2.715	3.3%
3	Methyl Parathion	3.320	2.8%
4	Parathion	4.291	2.8%
5	Ethion	7.478	2.0%
6	Trithion	7.973	2.1%
7	EPN	9.475	2.0%
8	Co-Ral	12.225	7.8%

Besides improving the analysis time, programming increased the minimum detectability of the last three peaks 5, 5 and 3 fold respectively.

The Aerograph Phosphorus Detector has been used to demonstrate advanced techniques for residue analysis for phosphorus-containing compounds. The Phosphorus Detector now makes it possible to simultanecusly analyze with two detectors - electron capture and phosphorus detector - this one group of pesticides by the powerful dual channel techriques (2). Dual channel chromatography of pesticides adds another dimension of quantitative and qualitative reliability for this analysis.

Acknowledgments

The assistance of Dudley M. Oaks for preparation of standards and technical advice is greatly appreciated. The assistance of Robert J. Thyken, Jr. as instrument operator is also appreciated.

References

1. Guiffrida, L., J.A.O.A.C., 47, 293 (1964).
2. Hartmann, C. H., Aerograph Research Notes, Summer 1966.
3. Karmen, A., Anal. Chem., 36, 1416 (1964).
4. Oaks, D. M., Dimick, K. P., Hartmann, C. H., Aerograph Publication W-122.

A Simple Evaporation Technique

by C. W. MILLER
University of Massachusetts
Cranberry Experiment Station, East Wareham, Massachusetts

Concentration of an extract to a small volume is often necessary prior to spotting on thin layer plates. In many instances,
the extract is concentrated to a desired volume, transferred to a
suitable container and final evaporation accomplished by a stream
of air or nitrogen. A known volume of solvent is then added and the
sample is ready for spotting. Often the vessel used for the final
evaporation step is not suitable as a storage vessel and the sample
must be transferred again. The following described method is considerably more rapid than air drying, it eliminates the necessity of
transferring the sample to a storage vessel and may be accomplished
with a minimum of expensive equipment.

The sample is first concentrated to incipient dryness using a
Rinco rotary vacuum evaporator with a 24/40 ₮ spindle. The flask is
then removed, rinsed twice with 2 ml of a desired solvent and the
rinses transferred to a 5 ml capacity screw cap vial.

The open end of a straight ground joint, 6 inches long with a
24/40 ₮, is snuggly sealed with a rubber stopper. The vial is placed
in the ground joint with a pair of forceps and sufficient water added
to the joint to cover 1/3 the height of the vial. When the ground
joint is affixed to the evaporator, it may be tilted as much as 45°

169

and placed in a hot water bath. The complete arrangement of the components is illustrated in figure 1.

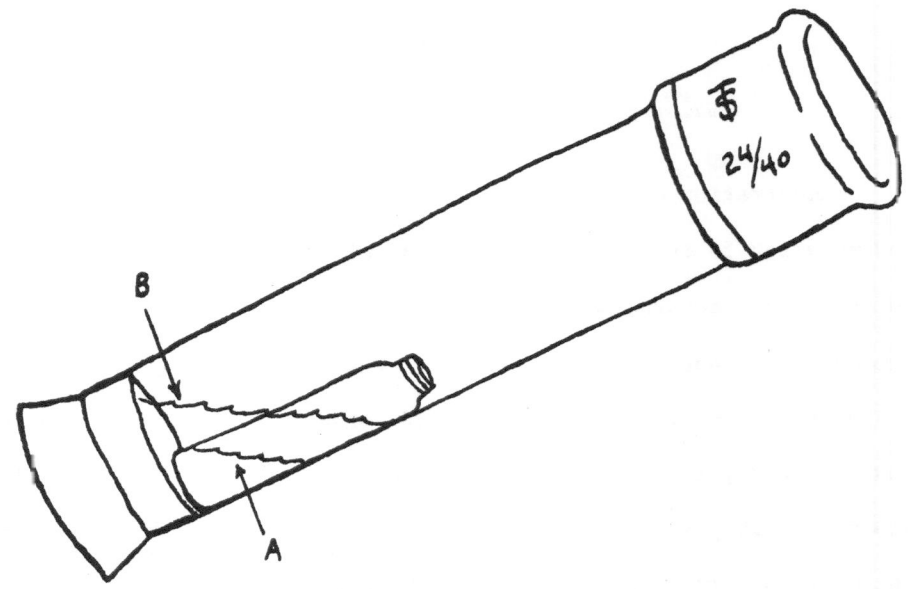

Figure 1. Orientation of components used in evaporation technique. (A) indicates level of solvent in vial, (B) indicates water level in ground joint.

Due to the screw threads of the vial being indented, the water will not enter. The vial will ride smoothly atop the rubber stopper as the joint rotates using the water as a lubricant. It has been found advantageous to use water from the water bath to fill the ground joint since temperature equilibrium is more rapid.

Once the solvent in the vial is reduced to a desired level, the original concentrating flask may be rinsed again and the entire process repeated as necessary to insure complete recovery of the compound of interest.

Care must be taken that excessive vacuum is not applied which may cause the water to boil over into the vial. Also, the vacuum must be released slowly so that the vial is not violently drawn up to the metal spindle of the evaporator.

The complete operation may be accomplished in 3-5 minutes, and the residue dissolved in any solvent to a maximum volume of 5 ml. By placing a teflon wafer within the screw cap, the vial may then be used to store the sample until ready for analysis.

Development of this technique supported in part by funds from Hatch Project 251.

Preparation and Identification of
2-Chloroethyl 1-Naphthyl Acetate

by T. E. ARCHER
Agricultural Toxicology and Residue Research Laboratory
University of California, Davis, California

In the past few years, increased interest in the physiological properties of naphthaleneacetic acid and its derivatives has been apparent. Guthrie (1) described the use of the methyl ester of naphthaleneacetic acid to inhibit the growth of potato tubers and to induce epinasty of tomato leaves. Mitchell and Stewart (2) studied the use of naphthaleneacetic acid to induce growth response in plants. Hesse and Davey (3, 4) experimented with naphthaleneacetic acid in the control of fruit drop of pear, apricot, and peach trees. Hartmann (5) employed naphthaleneacetic acid in a spray for thinning in olive trees to increase fruit size and provide for a normal crop set the following year.

Since naphthaleneacetic acid has been so widely used in agriculture, it has become increasingly important to have a sensitive analytical procedure for detecting it in fruits and leaves. Coulson et al. (6) have described a method based on gas chromatography and microcoulometric detection for chlorinated organic pesticides. Goodwin et al. (7) have detected chlorinated organic compounds by gas chromatography and electron affinity detection. Since these methods have an advantage of being extremely sensitive and have an increased specificity due to the halogen in the molecule, the analysis of 1-naphthaleneacetic acid as its chloro-ester

172

was undertaken. Success has been achieved in preliminary evaluation of the residue analysis for naphthaleneacetic acid as 2-chloroethyl 1-naphthyl acetate by gas chromatography in combination with the microcoulometric or electron affinity detectors.

The synthesis and identification of 2-chloroethyl 1-naphthyl acetate as a primary standard for analysis by gas chromatography is described below.

Experimental

Chemicals and Equipment. The 2-chloroethyl esters of short chain fatty acids have been prepared by Oette and Ahrens (8) for quantitative determination by gas-liquid chromatography. The preparation of 2-chloroethyl 1-naphthyl acetate was undertaken using this method with the following modifications:

Five grams of recrystallized α-Naphthaleneacetic acid (m. p. 130.5-131.5°C) was dissolved in 15 ml of redistilled 2-chloroethanol (b. p. 128.0-128.8°C) containing 10% (w/w) of boron trifluoride as described by Metcalfe and Schmitz (9). The solution was sealed in a 16 x 20 mm. Pyrex tube and heated for 20 minutes in a boiling water bath. The mixture was cooled in an ice bath and transferred to a 250 ml separatory funnel with three rinses of 33 ml of distilled water and several small rinses of petroleum ether. The water layer was extracted with three 100 ml portions of petroleum ether. The pooled petroleum ether extract was washed with 100 ml of distilled water, dried over anhydrous sodium sulfate, quantitatively filtered, and evaporated free of solvent under a warm air stream. The crude yield of ester was 93.7%.

Results and Discussion

The ester was purified by vacuum distillation to provide a clear, straw-colored liquid, b. p. 167°C (2.5 mm.), N^{20} D 1.5922.

Anal. Calcd. for $C_{14}H_{13}O_2Cl$: C, 67.59; H, 5.27; Cl, 14.26. Found: C, 67.68; H, 5.28; Cl, 14.12.

The infrared and ultraviolet spectra were consistent with the proposed structure.

2-Chloroethyl 1-naphthyl acetate analysed by gas chromatography and microcoulometry gave a single peak with a retention time of 2.7 minutes at 245°C and 100 ml per minute of nitrogen carrier gas. A 6-foot 1/4-inch O.D. stainless steel column was packed with 30/60 mesh acid washed Chromosorb R coated with 20% Dow 11 silicone grease.

2-Chloroethyl 1-naphthyl acetate analyzed by gas chromatography and electron affinity detector gave a single peak with a retention time of 3.5 minutes at 200°C and 80 ml per minute of nitrogen carrier gas. An 8-foot 1/8-inch O.D. stainless steel column was packed with 60/80 mesh Chromosorb W (HMDS) coated with a mixture of 5% Dow 710 silicone fluid and 5% SE-30 gum rubber.

References

1. J. D. GUTHRIE, Contrib. Boyce Thompson Inst., 10, 325 (1939).
2. J. W. MITCHELL and W. S. STEWART, Bot. Gas., 101, 410 (1939).
3. C. O. HESSE and A. E. DAVEY, Amer. Soc. Hort. Sci., 40, 55 (1942).
4. A. E. DAVEY and C. O. HESSE, Proc. Amer. Soc. Hort. Sci., 40, 49 (1943).
5. H. T. HARTMANN, Calif. Agric., 6, 7 (1952).
6. D. M. COULSON, L. A. CAVANAGH, J. E. DE VRIES, and BARBARA WALTHER, J. Agr. Food Chem., 8, 399 (1960).
7. E. S. GOODWIN, R. GOULDEN, and J. G. REYNOLDS, The Analyst, 82, 697 (1961).
8. KURT OETTE and E. H. AHRENS, JR., Anal. Chem., 33, 1847 (1961).
9. L. D. METCALFE and A. A. SCHMITZ, Anal. Chem. 33, 363 (1961).

Rapid Thin-layer Chromatographic Screening for Parathion in Canned Peaches Without Cleanup

by Daniel E. Ott, Fred E. Hearth, and Francis A. Gunther
Department of Entomology
University of California, Riverside, California

The extensive use of parathion as an insecticide on commercially grown peaches in California has created a need for a rapid screening method for the insecticide in the marketable crop. Using the method of KOVACS (1), with only slight modifications with respect to activation time of thin-layer chromatography (TLC) plates and the use of a simplified developing tank, concentrates of extractives of commercially canned peaches in methylene chloride stripping solution may be chromatographed and subjected to a series of chromogenic spray reactions to yield in about one hour qualitative and semi-quantitative information in the range 0.5 to 11 p.p.m. of residual parathion.

This procedure could serve as a cleanup step prior to final quantitative determinations by such analytical methods as oscillopolarography by the procedure of HEARTH et al. (2), automated analysis after the manner of OTT and GUNTHER (3), or gas chromatography according to the procedure of GIUFFRIDA and IVES (4).

Screening procedures for other pesticide residues in other substrates involving TLC without prior cleanup may be found in the reports by MORLEY and CHIBA (5,6) and OTT and GUNTHER (7).

A rapid automated screening procedure for technical grade

175

parathion in canned peaches has been reported by OTT and GUNTHER (8).

Method

Reagents

(a) TLC adsorbent.--Aluminum Oxide G. Warner-Chilcott Laboratories (formerly Research Specialties Co.), Richmond, Calif.

(b) Immobile phase.--15% N,N-dimethylformamide in diethyl ether.

(c) Developing solvent.--Methylcyclohexane (practical grade). Eastman Kodak Co., Rochester, N. Y.

(d) Chromogenic reagents.--(i) Dye solution: dissolve 100 mg. of tetrabromophenolphthalein ethyl ester (Eastman Kodak Co.) in 50 ml. of redistilled acetone. (ii) Silver nitrate solution: dissolve 0.5 g. of silver nitrate in 25 ml. of distilled water, dilute to 100 ml. with acetone. (iii) Citric acid solution: dissolve 5 g. of citric acid in 50 ml. of distilled water, dilute to 100 ml. with acetone.

Procedure

Preparation of TLC plates.--Prepare TLC plates 0.5-mm. thick in the usual manner from a slurry of 30 g. of aluminum oxide G in 35 ml. of distilled water. Allow the plates to air dry approximately 15 minutes then activate at 110° C. for 30 minutes. Plates may be used immediately upon cooling or stored in a dry place for future use.

Sample preparation and TLC.--Concentrate 20 ml. of a methylene chloride stripping solution of finely ground drained canned peaches

(1 g./2 ml.) to approximately 1 ml. in a Kuderna-Danish evaporative concentrator (9) over a steam bath. Finally concentrate to a volume suitable for TLC application (100 μl. or less; equivalent to extractives from 10 g. of peaches) with a gentle jet of clean, dry air while warming the tube to about 40° C. in a water bath. Vertically score the TLC plate into channels for each spot to be developed and prepare a hole in the adsorbent at the origin of the spot to prevent "flaking" as discussed by CHIBA and MORLEY (6) and by HEARTH et al. (2). Horizontally score the plate 10 cm. above the origin. Spot the entire concentrate on the TLC plate with a 100-μl. syringe or disposable micropipette. [Application time for many samples is considerably shortened by the use of one or more multiple-spotting devices as described by OTT and GUNTHER (3).] Also, spot on the same plate an appropriate control, a fortified control, and a standard. Prepare a fortified control by first spotting the desired amount of parathion in n-hexane solution followed by overspotting of a control concentrate, or, more realistically, a parathion standard is added directly to a control stripping solution prior to concentration.

Expose the spotted and air dried plate 20 to 30 minutes to the vapors only from the immobile phase in a closed, filter paper-lined TLC development tank: support the plate out of direct contact with the solvent mixture or place it in the empty side of a divided-bottom tank. This procedure permits extremely uniform saturation of the adsorbent with the immobile phase and was shown to give superior results to either the dipping procedure outlined by KOVACS

(1) or an alternate method of spraying the prepared plate with the immobile solvent prior to immersion in methylcyclohexane. At the end of the vapor saturation period, carefully introduce the developing solvent to a depth of about 0.5 cm. into the empty side of the tank or place the plate in a separate filter paper-lined tank containing the developing solvent and develop to the previously scored 10-cm. line.

Spray the air dried plate uniformly with the dye solution; the entire plate turns bright blue. Follow by a spray of the silver nitrate solution, wait 2 minutes, and finally spray the plate with the citric acid solution. An immediate background color change to yellow is produced with the last spray, while the R_f area containing parathion becomes lavender. Extractives from 10 g. or less of control peaches per spot yield no detectable blue or lavender areas.

KOVACS (1) has reported that the spot color produced with this insecticide is vivid blue and fades irreversibly after 30 to 40 minutes. However, we have found that color may be restored with a fine spray of water as much as 24 hours later.

Results and Discussion

Rapid comparative identification of parathion in fortified canned peaches from a lower limit of 0.5 p.p.m. up to 11 p.p.m. is possible by this method, which gives no discernible chromogen-reactive background in unfortified controls. This reported upper limit is not an absolute maximum but rather the highest fortification examined.

Parathion standards exhibit R_f values in the range of 0.70 to

0.76 while spots from fortified controls range from R_f 0.65 to R_f 0.74. This slight lag caused by the presence of peach extractives requires the inclusion of appropriately placed fortified controls on the same plate with test samples.

The successful application of this method for the detection of parathion residues in peaches strongly suggests the use of the method for rapid screening for Diazinon, malathion, parathion, Systox (thiono), Trithion, and other sulfur-containing organophosphorus insecticide residues in other crop substrates. TLC identification of these insecticides in the presence of extractives from strawberries, apples, and certain vegetable crops has already been reported by KOVACS (1).

Summary

Rapid screening for parathion above 0.5 p.p.m. in canned peaches is possible by thin-layer chromatography including a selected colorimetric spray detection system which does not respond to the natural components in the substrate. Approximately one hour is required from spotting of plate to interpretation of results.

References

(1) M. F. KOVACS, Jr., J. Assoc. Official Agr. Chemists 47, 1097 (1964).

(2) F. E. HEARTH, D. E. OTT, and F. A. GUNTHER, In preparation for ibid.

(3) D. E. OTT and F. A. GUNTHER. In press, ibid. (1966).

(4) L. GIUFFRIDA and F. IVES, *ibid.* 47, 1112 (1964).

(5) H. V. MORLEY and M. CHIBA, *ibid.* 47, 306 (1964).

(6) M. CHIBA and H. V. MORLEY, *ibid.* 47, 667 (1964).

(7) D. E. OTT and F. A. GUNTHER, In preparation for Bull. Environ. Contamination and Toxicol.

(8) D. E. OTT and F. A. GUNTHER, In press, J. Assoc. Official Agr. Chemists (1966).

(9) F. A. GUNTHER and R. C. BLINN, "Analysis of Insecticides and Acaricides," pp. 231-3 (1955), Interscience-Wiley, New York.

Paper No.   , University of California Citrus Research Center and Agricultural Experiment Station, Riverside, California.

Bulletin

Contents

AIMS AND SCOPE

The Bulletin of Environmental Contamination and Toxicology will provide rapid publication of significant advances and discoveries in the fields of pesticide residue research, air, soil, and water contamination and pollution, methodology, and other disciplines concerned with the introduction, presence, and effects of toxicants in the total environment.

Results of current research will be presented as brief reports providing information which is potentially useful to all individuals concerned with environmental contamination.

The articles will be free from restrictions imposed by purely scientific journals, particularly with respect to completeness of the studies reported and the attendant delays in publication.

Descriptions of new methods, procedures, or techniques shall be sufficiently detailed so as to permit direct application in other laboratories.

Review articles and obvious abstracts of papers forthcoming in other publications are not invited and probably will not be acceptable.

Articles suitable for inclusion shall be relatively short (less than 2,000 words) and will be prepared following specific instructions to permit reproduction by the photo-offset process from the original manuscript.

It is the hope of the Editorial Board that this Bulletin will provide a meeting ground for researchers who daily encounter problems related to the contamination of our environment and who welcome opportunities to share in new discoveries as they occur.

The Bulletin will be issued six times a year. This will be raised to 12 issues annually as demand increases.

Published bi-monthly by SPRINGER-VERLAG NEW YORK INC., *175 Fifth Avenue, New York, N. Y. 10010, Telephone (212) 673-9797. Six issues per year. Subscription price: $15 per year for institutions, $7.50 per year for individuals.*

Determination of Kelthane and Its Dichlorobenzophenone Degradation Product by Thin-Layer Chromatography and Oscillopolarography

by Daniel E. Ott, Fred E. Hearth, and Francis A. Gunther
Department of Entomology
University of California, Riverside, California

In a search (1) for an analytical method to distinguish quantitatively between the o,p'- and p,p'-isomers of Kelthane, an unsuccessful attempt was made to do so with thin-layer chromatography (TLC) and oscillopolarography. However, it has now been found that this combination of analytical techniques will separate and quantitate p,p'-Kelthane and a gas chromatographic degradation product (2,3), p,p'-dichlorobenzophenone.[1]

Method

Reagents

(a) TLC adsorbent.--Silica gel GF_{254} (manufactured by E. Merck, Darmstadt, West Germany; distributed by Brinkmann Instruments, Inc., Westbury, N. Y.).

(b) Electrolyte solution.--Prepare a 0.2 M aqueous solution of tetramethyl ammonium bromide (Eastman Organic Chemicals, Rochester, N.Y.); this solution is stable under normal laboratory

---

[1] The use of Kelthane and of dichlorobenzophenone hereafter applies to the p,p'-isomers. Kelthane is 1,1-bis(p-chlorophenyl)-2,2,2-trichloroethanol. NOTE: No appreciable residue metabolism or other degradation of Kelthane on field treated citrus was found by GUNTHER et al. (4).

Bulletin of Environmental Contamination & Toxicology,
Vol. 1, No. 5, 1966, published by Springer-Verlag New York Inc.

conditions.

(c) TLC chromogenic reagent.--Prepare a solution of 100 mg. of N,N-dimethyl-p-phenylazoaniline (methyl yellow) in 100 ml. of 95% ethanol. This solution retains its effectiveness for more than one month without special storage.

(d) Standard solutions.--Prepare a 100 µ g./ml. solution of Kelthane; weigh 5 mg. of the purified compound into a 50-ml. glass-stoppered volumetric flask, dissolve in, and dilute to the mark with, redistilled n-hexane. Similarly prepare a 100 µ g./ml. solution of dichlorobenzophenone in methylene chloride.

Special Apparatus

Davis Southern Analytical Differential Cathode Ray Polaro-trace Model A.1660A and Gajan oscillopolarographic cells (Western Scientific Associates, San Ramon, Calif.). Operate with an amalgamated #22 gauge silver wire as reference electrode (5).

Procedure

Prepare 0.5-mm. thick plates in the usual manner from a 1:2 slurry of silica gel GF$_{254}$ and distilled water. Air dry the plates about 15 minutes before activating them at 110° C. for 30 minutes. Spot samples of Kelthane or of dichlorobenzophenone or mixtures of the two in the range 0.5 to 20 µ g. by use of a micro-liter syringe. Shorten spotting time with a current of cool air blowing across the plate. Develop the plate in 10% methyl alcohol in n-pentane, in a closed TLC development tank lined with filter paper, to a line previously scored 10 cm. above the applied spots.

Examine the developed and cool air-dried plate under an ultraviolet lamp at 2537 Å and mark the visible spots.

For additional qualitative information spray half of a developed TLC plate with the chromogenic reagent and save unsprayed the mirror-image half for oscillopolarographic analysis as described below.

Scrape uniform areas of the plate which encompass each spot into separate 12-ml. centrifuge tubes. Add 1.0 ml. of 95% ethyl alcohol to each and mix thoroughly with a stirring rod for 30 seconds. Add 1 ml. of electrolyte solution, mix briefly by agitation, and centrifuge. Decant the supernatant solution into a Gajan oscillopolarographic cell. Analyze the solution oscillopolarographically after deoxygenation by bubbling prepurified water-pumped nitrogen through it for 3 minutes; scan from -0.5 to -1.0 v. to observe the two peaks from Kelthane and from -1.0 to -1.5 v. for the dichlorobenzophenone peak.

Results and Discussion

R_f values for 5 μg. each of Kelthane and dichlorobenzophenone chromatographed in adjacent channels are 0.29 to 0.36 and 0.77 to 0.86, respectively, when the leading and trailing edges of each spot are measured. R_f values are essentially the same when the two compounds are chromatographed in admixture. Variations in R_f values are no more than ± 0.03 for either compound in the range 0.5 to 10 μg. whether chromatographed singly or in admixture. It should be noted, however, the higher R_f values are typically obtained when TLC plates of decreased thickness are used or when the ambient temperature is increased above 25° C. Applications of 5 µg. quantities of Kelthane and dichlorobenzophenone to TLC plates 0.25 mm. thick show R_f values of 0.36 to 0.43 and 0.82 to 0.39, respectively.

In addition to detection of Kelthane and dichlorobenzophenone by the use of fluorescent quenching of fluorescent TLC plates under ultraviolet light, a sensitive chromogenic reagent, N,N-dimethyl-p-phenylazoaniline (methyl yellow), may also be used. Use of this spray reagent for the detection of Kelthane and other chlorinated hydrocarbons on paper chromatograms has been reported by KRZEMINSKI and LANDMANN (6). We have found that exposure to strong ultraviolet light as recommended (6) is not essential to obtain good detection on TLC plates of both Kelthane and dichlorobenzophenone. Best detection of the purified compounds at levels of 0.5 and 1.0 µg. is obtained approximately 10 minutes after application of the chromogenic reagent. As the plates are air dried under a draft of cool air, the adsorbed compounds become visible as deep yellow spots against a lighter yellow background, but the color distinction between background and spots is short lived (about 15 minutes) because of increasing intensity of background color upon continued exposure to ordinary light. For this reason, R_f areas should be marked as soon as spots become visible. When plates are exposed to strong ultraviolet radiation immediately after spraying, background color intensifies and obscures R_f areas of dichlorobenzo-

phenone at any level and for Kelthane at 1.0 µg. or less. At higher µg. levels, the Kelthane complex becomes a deep orange color
which is relatively stable. When Kelthane and dichlorobenzophenone
are chromatographed in the presence of orange peel extractives at
a fortification level of 1 p.p.m. or less, detection of the test
compounds by the use of methyl yellow indicator is no longer
possible.

Oscillopolarography of purified Kelthane at 2.5 µg./ml. in the
analytical solution reveals two reduction potentials: -0.65 and
-0.85 v. These reduction potentials become slightly more negative
with increasing concentration from 1.3 to 10.0 µg./ml. in the final analytical solution. The single reduction potential for purified dichlorobenzophenone was typically -1.26 ± 0.01 v. over the
range 2.5 to 10 µg./ml. Both compounds in these concentration
ranges exhibited straight-line relationships between relative
polarographic units (relative to maximum instrumental sensitivity)
and concentration (µg./ml.) in the final analytical solution. Reduction potentials for compounds from standard solutions compared
to those following TLC showed no significant variation.

It should be pointed out that the reduction potentials of the
two compounds of interest are sufficiently separated that it
should be possible to estimate them in admixture without the need
for TLC separation and isolation. However, in unknowns from formulations or crop samples the TLC procedure would be needed as the
cleanup (or part of the cleanup) method. In this situation, for

quantitative purposes unknowns must be related to standards chromatographed on the same plate at approximately the same amounts to account for recovery losses in the TLC procedure. These losses may vary from plate to plate, but are adequately consistent for any given plate; see the report by HEARTH et al. (5) for a detailed discussion of the same problem with the acaricide Morestan under TLC conditions.

Polarography of Kelthane has been briefly mentioned earlier (7). To our knowledge, however, the present report is the first mention of the polarographic analysis of p,p'-dichlorobenzophenone even though the polarography of ketones in general is well established.

References

(1) W. E. WESTLAKE, R. T. MURPHY, and F. A. GUNTHER, Bull. Environ. Contamination and Toxicol. 1(1), 29(1966).

(2) F. A. GUNTHER, J. H. BARKLEY, R. C. BLINN, and D. E. OTT, Pesticide Research Bull., Stanford Research Inst. 2(2), 3(1962).

(3) _____, in Adv. Pest Control Research 5, 308(1962), R. L. Metcalf, ed.

(4) _____, R. C. BLINN, L. R. JEPPSON, and J. H. BARKLEY, J. Agr. Food Chem. 5, 595(1957).

(5) F. E. HEARTH, D. E. OTT, and F. A. GUNTHER, J. Assoc. Offic. Anal. Chemists, in press for August (1966).

(6) L. F. KRZEMINSKI and W. A. LANDMANN, J. Chromatog. 10, 515 (1963).

(7) E. J. GAJAN and J. LINK, J. Assoc. Offic. Agr. Chemists 47, 1119(1964).

Paper No. 1696, University of California Citrus Research Center and Agricultural Experiment Station, Riverside, Calif.

Rate of Transfer of DDT
from the Blood Compartment[1,2]

by J. M. Witt[3,4], W. H. Brown[4], G. I. Shaw[4], L. S. Maynard[3],
L. M. Sullivan[4], F. M. Whiting[4], and J. W. Stull[4]

A number of workers have analyzed blood for pesticides

(1-10) and some have correlated the level of pesticides in

blood with that in body fat (2, 5). These correlations were

obtained with the intention of utilizing blood samples in

lieu of adipose tissue biopsy samples for the purpose of

determining pesticide storage levels in various segments of

the population. Basic to any interpretation of such a cor-

relation is an understanding of the rate of transfer of a

pesticide from the blood following a discrete accumulation.

This rate of loss must be known in order to interpret whether

the pesticides found in the blood are a reflection of an

immediately prior overt intake or a reflection of an equilib-

rium with pesticides stored in adipose tissue.

[1]Arizona Agricultural Experiment Station Technical Paper No. 1134

[2]This work was supported in part by Grant No. EF-00627-02 from
the U.S.P.H.S. and a Grant from the United Dairymen of Arizona

[3]Department of Entomology, The University of Arizona, Tucson

[4]Department of Dairy Science, The University of Arizona, Tucson

187

Workers who have reported results of the analysis of various kinds of blood for pesticide residues have dealt with the level of pesticide in blood in equilibrium with a certain pesticide exposure history rather than the rate of change of level of pesticide in blood following exposure. Basic to the detection of pesticides in blood is the complete extraction of the pesticide from the blood. Since several authors have used different extraction methods, but offer little evidence as to whether the extraction was complete, it is important to establish that the methods used are adequate. Methods which have been used include KOH hydrolysis followed by pentane extraction or extraction with ethyl ether/ethyl alcohol, ethyl ether/acetone, ethyl ether only, n-hexane only, and petroleum ether only. Crosby and associates (1,5) extracted blood for DDT by direct hydrolysis of the blood with KOH and extraction of the hydrolysate with pentane. This method dehydrochlorinates DDT to DDE and therefore prevents the simultaneous detection of the two products and also precludes a simple detection of the amount of lipid present. These authors generally found good correlation (0.95) between the amount of pesticide in the blood and the amount in the adipose tissue for cows dosed at 30 to 600 ppm with DDT in the diet. If one assumes that the blood contained 400 mgm of lipid per 100 g. of blood, there was about 1/3 the amount of DDT and DDE in the blood as in the adipose tissue when compared on an extractable lipid basis. Dale et al. (2)

extracted blood plasma only using ethyl alcohol and ethyl ether
(Bloor (11)), which has long been accepted as sufficient for
complete extraction of lipid. Since the chlorinated hydrocarbon insecticides are fat soluble compounds, the complete
extraction of lipid material is probably an important criter-
ion for the complete extraction of the insecticide. These
authors found about 10 times more DDT in the plasma lipid than
that found in adipose tissue lipid from rats receiving 200 ppm
of DDT in their diet. Jain et al. (4) used 3 successive
extractions with an acetone-ether mixture to extract insecticides from the blood of rats which had received an LD50 dose
of the insecticide. They showed that the recoveries were not
enhanced by use of a greater number of extractions. They
analyzed the extract directly without cleanup. This was
probably successful because very small samples could be used
since the levels of pesticide in the blood were quite high.
Dale et al. (3) extracted blood with n-hexane only and analyzed
the extract without cleanup. They noted that successive extractions did not increase the yield of insecticide, but also
noted that they apparently experienced low recoveries which
may have been due to binding with some constituent of the
serum. They showed levels of 0.01 ppm of DDE and 0.006 ppm
of DDT in the whole blood of persons without occupational
exposure. Earlier authors (7,9,10) used ether or ether and
acetone solvent systems to extract DDT from blood. They failed
to detect any DDT, probably due to the limits of sensitivity

of their detection methods. Stiff (8), using an ether/acetone solvent system and a total chloride detection method, reported rather large values (6.8 to 9.8 ppm) for DDT residues in whole blood from rats receiving a 550 mg/kg. oral dose. Stiff (8) reported on the appearance of DDT in the blood in relation to an oral dose, but no authors have reported on the rate of decline of an insecticide from blood when the decline rate was not compounded by an input rate.

It was the purpose of this study to determine the rate of transfer of DDT from the blood compartment following the establishment of a unique maximum level of pesticide in the blood by a single intravenous injection. A second purpose was to compare the amount of pesticide extracted using either an apolar solvent only or a polar solvent in combination with an apolar solvent and to relate this to the total amount of lipid extracted.

Materials and Methods

Two mature lactating Holstein cows were dosed with a preparation of DDT calculated to be approximately 4 ppm on the basis of consumed feed. The preparation was a mixture of two fractions: 1) 150 ml. of peanut oil, 150 ml. Atlox 1054-A and 15 g. p,p'DDT; 2) 1 g. sodium cholate, 12 g. lecithin, (soy refined) 8 g. NaCl and 685 ml. distilled water. Just before dosing 16 ml. of 1) and 34 ml. of 2) were combined and shaken vigorously. The appropriate amount was then injected directly into the jugular vein. Blood was collected into previously

heparinized bottles before injection and at 1-10 min. and 1,2,4,
8,16 and 24 hours post injection from the subcutaneous abdominal
vein.   The samples were frozen until time of analysis.

The blood was extracted by two methods.   The first consisted of weighing out 25 or 50 g. (dependant on expected
pesticide level) into a 250 ml. beaker.   The blood was then
poured into a 500 ml. separatory funnel which contained 3 ml.
of 5% potassium oxalate.   The beaker was rinsed into the
funnel with two 10 ml. washings of distilled water.   Fifty ml.
of absolute ethanol were then added following which the separatory funnel was inverted 4 times.   One hundred ml. of ethyl
ether were then added followed by 1 min. of shaking after
which 50 ml. of pentane were added and it was again shaken
for 1 min.   Distilled water was then added until the funnel
was ca 3/4 full.   The funnel was then inverted 6 times and
allowed to stand 10 min. after which the lower layer was discarded.   This step of adding water and discarding was repeated
twice.   The solvent layer was then poured through a funnel
containing sodium sulfate and a glass wool plug into a 250 ml.
beaker.   The separatory funnel was rinsed through the funnel
with 4 washings of 10 ml. of pentane each.   The solvent was
then evaporated to near dryness and transfered from the beaker
to a 15 ml. centrifuge tube with pentane.   An aliquot representing 5 g. of blood was then removed for fat determination.
The contents of the centrifuge tube were then poured onto a
column pre-washed with the eluant containing 4 in. of activated

(130°C,16 hr.) florisil and 1/2 in. of sodium sulfate. The centrifuge tube was washed twice with pentane. The pesticide was eluted with 200 ml. of 10% ether in pentane. The eluant was then evaporated to less than 5 ml. and transferred to a 5 ml. centrifuge tube. The analysis for DDT was then carried out by electron capture gas chromatography as previously described (12). The second method of extraction from the blood was identical with the first except that the alcohol and potassium oxalate steps were omitted.

The aliquot for quantitative blood fat determination was evaporated to near dryness at room temperature and then the balance of the solvent was removed by leaving it in an oven for 1 hr. at 50-60°C. Blood lipid was then determined by direct weighing.

Results and Discussion

The results are shown in Table 1. It is important to note that when no alcohol is used in the solvent system approximately 10% as much lipid is extracted as when the alcohol is included. If fat soluble pesticides are uniformly distributed in body lipids, then only 10% of the pesticide should also have been recovered, and the concentration when expressed in terms of ppm in lipid should have remained constant. However, the total pesticide recovered increased by a factor of about 1.6 instead of a factor of 10. This implies that about two-thirds of the pesticide is associated with the easily extractable lipid and the other one-third is associated with that

90% which is more difficult to extract. It is not clear whether

these two situations are distinct because of location, i.e. the

90% fraction is in the interior of the red blood cells, or

whether they are distinct because of binding or the chemical

nature of the lipids, i.e. they are phospholipids, sterol

esters, or other alcohol soluble lipids. The choice of extrac-

tion method would depend on whether one wished to simplify the

cleanup problem by failure to extract all the lipid, or wished

to determine the total pesticide content of the blood.

TABLE 1

Comparison of the Extractable Lipid and DDT from Blood
Using Two Solvent Systems

Blood Sample	Pentane + Ethyl Ether			Pentane + Ethyl Ether + Ethyl Alcohol		
	Mg fat /100 g. blood	PPM DDT blood lipid	whole blood	Mg fat /100 g. blood	PPM DDT blood lipid	whole blood
511-1	28	180	0.05	416	26.6	0.11
511-2	26	27	0.007	400	4.4	0.018
511-3	32	19	0.006	398	2.9	0.012
511-4	32	16	0.005	388	1.8	0.007
511-5	42	6.0	0.0025	426	0.55	0.0023
511-6	30	2.3	0.0007	396	0.092	0.0003

The results of the rate of decline study are shown in

Figure 1. The data do not cover a sufficient period of time

to clearly demonstrate whether this situation represents the

classical two compartment process. The rapid rate of decline

in the early part of the process has a half-life of 60-80 min.

and continues for 6 hr., the slower rate of decline in the

later part of the process has a half-life of about 8 hr. It

appears that equilibrium is nearly re-established 24 hr. after

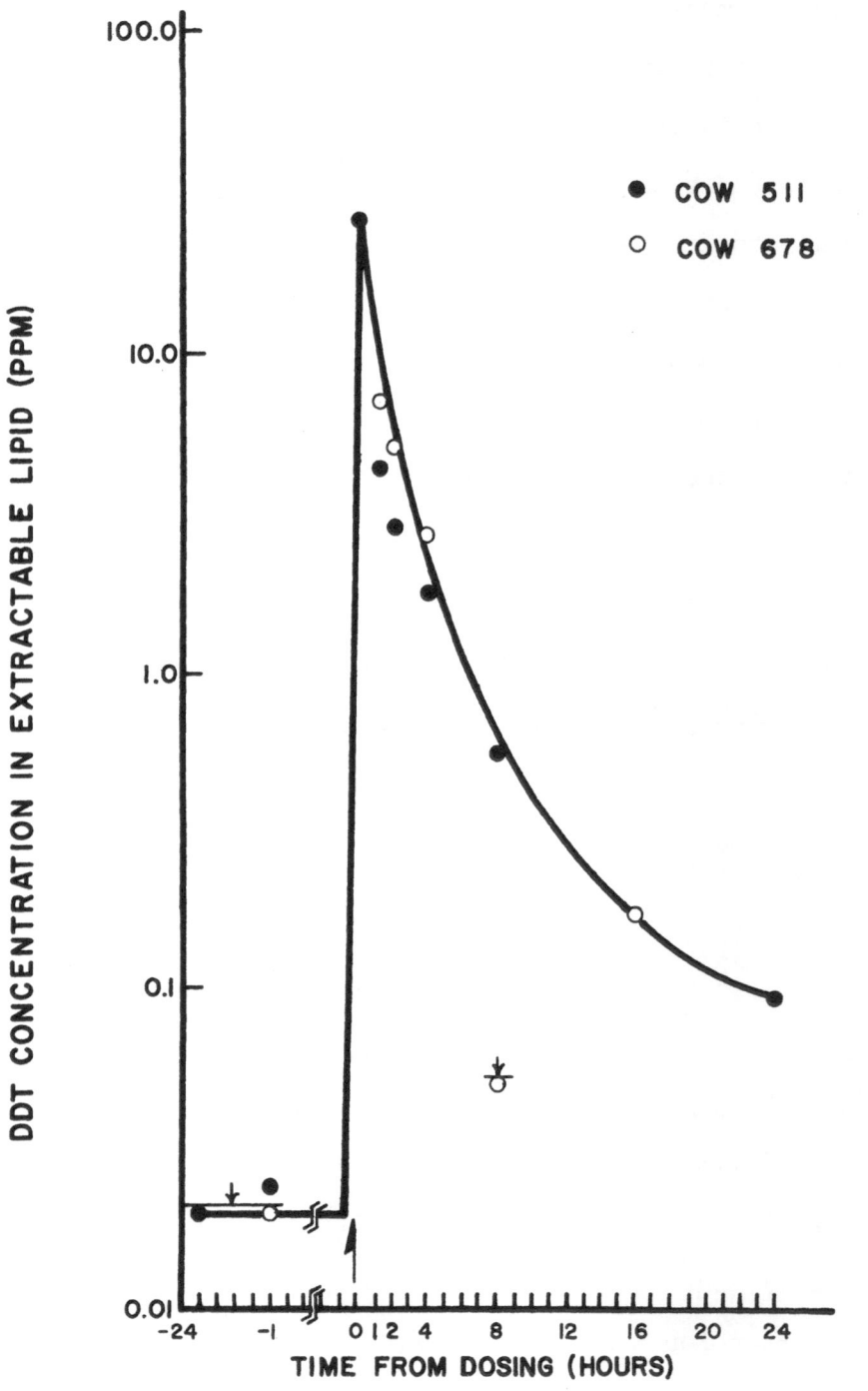

FIG I. RATE OF DECLINE OF DDT FROM BLOOD
AFTER INTRAVENOUS DOSING

cessation of pesticide absorption into the blood. It is interesting to note that the level of DDT in the blood lipid of the cow which has not received any overt exposure to DDT is 0.02 ppm or less (0.00008 ppm in whole blood) which is considerably less than the 1.5 ppm normally found in human blood from persons with no occupational exposure. This is probably due to the fact that the diet of the cow contained less than 0.1 ppm of DDT and was storing less than 0.5 ppm of DDT and DDE in its milk fat. This low level in the normal cow's blood is near the reagent blank response and makes the monitoring of normal cows by this method much more difficult than monitoring occupationally unexposed people. The recovery standards for this study ranged from 71 to 137% with a mean value of 99.6%.

Calculation of the level of pesticide expected to be found in the blood can be based on the weight of the cow, blood volume, percent lipid, and the amount of DDT injected. On this basis, it was calculated that only 10.4% of the dose of DDT was accounted for in the "zero time" sample. However, the "zero time" sample was taken from 1 to 5 minutes after injection, so it is not clear whether this low value is due either to poor recovery or an extremely rapid transfer from the blood to other compartments.

Extremely rapid disappearance from the blood following i.v. injection is frequently found in the case of electrolytes. Such behavior is especially striking in the case of manganous ion where, in the human subjects, only about 2% of the dose per liter of blood (near 10% of the dose for the entire blood supply) is found within 5 min. after injection, and by one hour, about 0.15%/liter (13). In the case of manganese, it was suggested that rapid movement into the intracellular compartment accounted for this behavior.

Conclusions

The rate of decline of DDT from the blood has been shown to be quite rapid and recovery of pesticide from the blood has been related to the recovery of lipid from the blood.

References

1. D. G. Crosby and T. E. Archer, Bull. Evniron. Contam. and Tox. 1, 16 (1966)

2. W. E. Dale, T. B. Gaines, and W. J. Hayes, Jr., Tox. and Appl. Pharm. 4, 89 (1962)

3. W. E. Dale, A. Curley, and C. Cueto, Jr., Life Sci. 5, 47 (1966)

4. N. C. Jain, C. R. Fontan, and P. L. Kirk, J. Pharm. Pharmacol. 17, 363 (1965)

5. R. C. Laben, T. E. Archer, D. G. Crosby and S. A. Peoples, J. Dairy Sci. 48, 701 (1965)

6. J. D. Judah, Brit. J. Pharmacol. 4, 120 (1949)

7. M. I. Smith and E. F. Stohlman, Pub. Health Rep., U.S.P.H.S. 59, 984 (1944)

8. H. A. Stiff, Jr. and J. C. Castillo, J. Biol. Chem. 159, 545 (1945)

9. E. P. Laug, J. Pharmacol. $\underline{86}$, 324 (1946)

10. E. P. Laug, J. Pharmacol. $\underline{86}$, 332 (1946)

11. W. R. Bloor, J. Biol. Chem. $\underline{17}$, 377 (1914)

12. J. M. Witt, F. M. Whiting, W. H. Brown, and J. W. Stull, J. Dairy Sci. $\underline{49}$, In Press (1966)

13. L. S. Maynard and G. C. Cotzias, J. Biol. Chem. $\underline{214}$, 489 (1955)

Acetylcholinesterase Levels in
Brains of Fishes from Polluted Waters

by A. K. Williams and Carl R. Sova
Southeast Water Laboratory
Department of the Interior, Athens, Georgia

One of the acute toxic effects of organophosphorous compounds is due to their ability to severely inhibit the enzyme,
acetylcholinesterase. This inhibition has been demonstrated
many times both in vitro and in vivo. Studies by Weiss (1,2,3)
have shown that fish exposed to various sublethal and lethal
concentrations of several organophosphorous compounds exhibited
a reduced level of AChE activity in the excised brain tissue.
Weiss has also shown that death of several species of freshwater
fishes occurs when brain AChE activity is 40 - 70% inhibited when
compared to that of nonexposed control fish of the same species
(1). This author has also suggested that brain AChE levels of
fish could be utilized as a means of detecting the presence of
organophosphorous pollutants in natural waters (1,2,3).

The Ashley River, Charleston, South Carolina, was chosen
as a study site because sporadic fish kills had been occurring

198

since 1961 in an area which receives wastes from a plant which manufactures a variety of organophosphorous compounds, including pesticides and defoliants.

Experimental

Moribund and apparently healthy (live) Atlantic menhaden, Brevoortia tyrannus (Latrobe), and apparently healthy (live) Atlantic croakers, Micropogen undulatus (Linnaeus), were collected from the same general area of the Ashley River. Menhaden, for control purposes, were collected by net from the off-shore waters in the vicinity of Charleston, South Carolina. Similarly, control croakers were taken by net from the Ashapoo River in South Carolina. Upon collection, all fish were frozen on dry ice, promptly shipped to the laboratory, and stored at -18 C prior to analysis. The collections were made in late June and early July, 1965.

The intact brain was removed from the frozen fish and weighed on tared aluminum foil. The entire brain was then homogenized in 2.0 ml of 0.2 M phosphate buffer, pH 8.2 containing 0.2 M NaCl and 0.002 M $MgCl_2$ using a glass tissue homogenizer. The resulting homogenate was diluted with buffer to a final concentration of 10 mg brain/ml. This stock brei was then diluted with buffer to obtain the desired concentration for assay. Reactions were carried out using 5.0 mg brain tissue (0.5 ml of stock brei), 0.5 ml of buffer, and 4.0 μM of acetylcholine chloride in a total volume of 2.0 ml. Reactions were allowed to incubate 20 minutes at 25 C. Residual acetylcholine was determined by the alkaline hydroxylamine - $FeCl_3$ method of Hestrin (4). The reactions were clarified by brief centrifugation immediately after color development. This entire procedure is essentially that of Weiss (1) with minor modifications. The results of these experiments are shown in Table 1.

TABLE 1

Acetylcholinesterase Activity of Fish Brain Homogenates

Source	Number Assayed	Specific Activity* Extremes	Mean	Percent Inhibition
Control Menhaden	18	0.71-1.49	1.09	----
Moribund Menhaden Ashley River	8	0.48-0.67	0.58	46.8
Live Menhaden Ashley River	8	0.72-1.14	0.91	16.5
Control Croakers Ashapoo River	8	1.02-2.34	1.48	----
Live Croakers Ashley River	8	0.72-1.13	0.95	35.8

*Specific Activity = μm acetylcholine hydrolyzed/mg. brain tissue/hr. Reactions were incubated at 25 C for 20 min.

Menhaden taken from the Ashley River in a distressed condition showed a 46.8% inhibition of brain AChE when compared to the control group; while menhaden collected from the same source, showing no obvious sign of distress, were found to be 16.5% inhibited. Although moribund croakers were not observed, croakers taken from the Ashley River were found to have 35.8% less AChE activity than the controls.

Although the specific activity values of some of the Ashley River fishes fall within the range of values of the control fishes, statistical analysis has established that the two groups are, in all cases, significantly different.

The statistical significance of the mean AChE activity values shown in Table 1 was determined with a "t-test" (1). For menhaden, the null hypothesis that the difference between two population means was zero was tested for each of the following cases: control minus live; control minus moribund; and live minus moribund. Similarly, for croakers, the following was tested: control minus live.

All "t" values were greater than the 0.05 probability of a larger value. Thus, the hypothesis was rejected and the following conclusions were drawn about specific AChE activity in brain tissue: The activity of menhaden collected from off-shore areas was greater than both live and moribund menhaden from the Ashley River; in the Ashley River, the activity of live menhaden was greater than that of moribund menhaden; and the activity of croakers from the Ashapoo River was greater than that of croakers from the Ashley River.

In an effort to locate possible sources of organophosphorous pollution, water samples were taken from several industrial outfalls along the Ashley River, including those of the pesticide and defoliant manufacturing plant. These water samples were extracted with chloroform, and the chloroform fraction was adjusted to a volume such that 2.5 ml chloroform was equivalent to 1.0 ml of water. The chloroform fractions were tested for the presence of AChE inhibiting materials by placing a known quantity of chloroform extract in a test tube and evaporating in air stream. Approximately 3 units of purified bovine erythrocyte AChE was then added to the tubes and allowed to incubate for 30 minutes at 35 C. Acetylcholine was added and the enzyme activity determined as previously described. From a total of 12 samples assayed, only 3 exhibited anti-AChE activity when compared to control reactions (Table 2). It is significant that samples one and two were taken from effluents of the pesticide plant and sample three was obtained from a fertilizer plant.

Analysis of the water samples by gas chromatographic techniques indicated the presence of at least two anti-AChE compounds. Subsequent analysis of these materials by infra-red, nuclear magnetic resonance, and mass spectroscopy confirmed the identity of these compounds as 0,0 diethyl-0(2,4-dichlorophenyl) phosphorothioate and S,S,S-tributyl phosphoro-trithioate (5).

TABLE 2

In Vitro Inhibition of AChE in Extracts of Waste Water
Being Discharged into the Ashley River

Sample Number	ml of Chloroform Extract Added	Percent Inhibition
1 [1]	1.25	95
1 [1]	0.10	30
2 [1]	1.25	42
3 [2]	1.25	29
4-12 [3]	1.25	none

[1] Collected from the waste effluent of the pesticide plant.

[2] Collected from the waste effluent of the fertilizer plant.

[3] Collected from other waste sources.

Discussion

The moribund menhaden collected from this area exhibited a
degree of AChE inhibition which has been shown to result in the
death of certain species of freshwater fishes (Weiss 1,2,3).
Furthermore, apparently healthy croakers collected from the
Ashley River exhibited AChE inhibition which approached that
thought to be critical for the more sensitive species of fresh-
water fishes (Weiss 1,2,3). While we have not shown that the
extensive fish kills which have occurred in the Ashley River over

203

a period of years are due solely to the inhibition of AChE, it is our opinion that these fish kills are due, at least in part, to the inhibition of this enzyme. The data also indicate, as Weiss (1,2,3) suggested, that AChE inhibition in fish-brain tissue in conjunction with chromatographic or chemical analysis has considerable potential as a means of monitoring waters for the presence of organophosphorous compounds.

Summary

Distressed menhaden collected from the Ashley River, South Carolina, were found to have 46.8% less acetylcholinesterase (AChE) activity in brain homogenates as compared to menhaden collected from offshore waters. Menhaden and croakers also taken from the Ashley River, but not in a distressed condition, were found to be 16.5 and 35.8% inhibited, respectively. AChE inhibiting materials were found in three of twelve waste water samples collected from the vicinity of the Ashley River.

References and Notes

1. Weiss, Charles M. Ecology, 39(2), 1958.

2. Weiss, Charles M. Sewage and Industrial Wastes, 31(5), 1959.

3. Weiss, Charles M. Trans. Am. Fish. Soc., 90(2), 1961.

4. Hestrin, S. J. Bio. Chem., 180, 1949.

5. Snedecor, G. W. Statistical Methods. 5th ed., Iowa State College Press, Ames, Iowa, xii, 534 pp.

6. Teasley, J. I. (Personal Communication).

Aflatoxin and Chromosomal Studies

by D. T. Cappucci, Jr.*

Animal Husbandry Department
University of California, Davis, California

Aspergillus flavus is known to be a common fungal
contaminant on groundnuts, groundnut cake, and meal.
Four metabolites have been designated as comprising
the toxins produced by the fungus. Collectively, the
B_1, B_2, G_1, and G_2 toxic factors have come to be called
aflatoxin; the B_1 form being considered the most common
naturally occurring form.

Since its discovery in a 1960 outbreak of "Turkey X"
disease,[1] aflatoxin has been shown to have carcinogenic
activity. Liver carcinomas have appeared in laboratory
rats[2,3] and trout[4] fed diets containing aflatoxin.
Comparative studies with other species - viz. pig[5],
cow[6], duck[7], chicken[8], monkey[9], and guinea pig[10]—
fed similar toxic diets, have revealed lesions of liver
cell damage and/or bile duct hyperplasia.

Theron postulated that aflatoxin has a direct
action on the liver cell membrane and membranes of
intracytoplasmic structures.[7] Juhasz and Greczi demonstrated with tissue cultures of calf kidney that
extracts of the toxin affect both the cellular nuclei
and cytoplasms.[11]

---

*Present address: 425 Vicksburg St., San Francisco,
Calif.,94114.

205

Bulletin of Environmental Contamination & Toxicology,
Vol. 1, No. 5, 1966, published by Springer-Verlag New York Inc.

Lilly, using roots of seedlings of <u>Vicia faba</u> (Sutton's prolific "Longpod"), demonstrated the induction of chromosomal aberrations by aflatoxin.[12] She observed a highly significant increase in abnormal anaphases of roots given treatments with aflatoxin. Significantly, most of the abnormalities consisted of chromosomal fragmentation with occasional bridges. Lilly[12] claims that Withers (personal communication) induced chromosomal breakage in human red blood cells in culture, using aflatoxin.

The present author proposes the hypothesis that aflatoxin's action may cause chromosomal aberration(s) under conditions of <u>in vivo</u> animal experimentation. Using animals given appreciable quantities of aflatoxin, he suggests that such studies may furnish evidence of chromosomal changes (e.g. via bone marrow cultures), besides the already cited pathologic lesions. However, the quantitative level(s) of aflatoxin necessary to induce chromosomal alterations, <u>in vivo</u>, may be higher than the amount(s) needed to produce tissue effects. In any event, no study appears to have been published giving the results of investigations of the action of aflatoxin on animal chromosomes.

References

1. F.D.Asplin and R.B.A.Carnaghan, Vet. Rec. 73, 1215 (1961).
2. M.C.Lancaster, F.P.Jenkins, and J.McL.Philp,Nature (Lond.)192, 1095 (1961).
3. J.M.Barnes and W.H.Butler, Nature (Lond.)202, 1016 (1964).
4. L.M.Ashley, J.E.Halver, and G.N.Wogan, Fed. Proc. 23, 105 (1964).
5. J.P.Raynaud, Rev. Elevage Med. Vet. Pays Trop. 16, 23 (1963).

6. D.Horrocks, A.W.A.Burt, D.C.Thomas, and M.C. Lancaster, Anim. Prod. 7, 253 (1965).

7. J.J.Theron, N.Liebenberg, and H.J.B.Joubert, Nature (Lond.) 206, 908 (1965).

8. H.I.Chute, S.L.Hollander, E.S.Barden, and D.C. O'Meara, Avian Diseases. 9, 57 (1965).

9. T.V.Madhavan, P.G.Tulpule, and C.Gopalan, Arch. Path. 79, 466 (1965).

10. W.H.Butler, J.Pathol. Bacteriol. 91, 277 (1966).

11. S.Juhasz and E.Greczi, Nature (Lond.) 203, 861 (1964).

12. L.J.Lilly, Nature (Lond.) 207, 433 (1965).

Gas Chromatographic Determination of Zinophos Residue in Soil[1]

by DAVID R. COAHRAN

Department of Agricultural Chemistry
Washington State University
Pullman, Washington

Although a spectrophotofluorometric method of analysis for

Zinophos ($\underline{0},\underline{0}$-diethyl $\underline{0}$,2-pyrazinyl phosphorothioate) is already

available(3), it seemed probable that a less complex procedure

could be developed, capitalizing on the great selectivity for

phosphorus compounds of a gas chromatograph equipped with a

sodium flame detector(1).

Chromatographic analysis proved feasible and is described

below.

Procedure

One hundred fifty grams of soil is placed in a medium

porosity 45 x 127 mm Alundum thimble. The thimble is loosely

plugged with absorbent cotton to prevent the soil from being

splashed out. The soil is extracted overnight in a 200 ml Soxhlet

extractor. The extract is evaporated nearly to dryness at

[1]Scientific paper 2857, College of Agriculture, Washington State
University. Work was conducted under project 1793.

reduced pressure in a rotary evaporator, transferred to a 10 ml volumetric flask and made up to volume.

A Wilkens Instrument and Research Model 204 Chromatograph fitted with a 1/8" x 5' glass column is used for analysis.

Both 5% Dow 11 on 60/80 mesh Chromosorb W and 5% Dow 200 on 80/90 mesh Anachrom ABS have been used as column packing. Both are satisfactory although the latter appears to give slightly sharper peaks. Column temperatures between 165° and 180° C are suitable with both packings. The carrier gas is prepurified nitrogen at a flow rate of about 30 ml/minute. A sodium flame detector(1) was used for the major part of this work, but a Varian-Aerograph (formerly Wilkens Instrument and Research, Inc.) phosphorus detector has also been used and is well adapted to this procedure. The sodium flame detector gives a limit of detection of about 0.06 p.p.m. Zinophos in the extract. In our hands the Aerograph phosphorus detector's limit is about 0.01 p.p.m. For 150 gram soil samples and a final extract volume of 10 ml as described, the calculated limits of detection in soil are .005 and .0007 p.p.m. respectively.

Discussion

Data on the recovery of Zinophos from several soils are given in Tables I and II. As can be seen from Table 1, recovery is good in all cases. Table II shows that the recovery of Zinophos from soil does not decrease greatly upon frozen storage.

None of the soils used shows any chromatographic components other than those attributable to organophosphorus pesticides.

Getzin(2) has used the sodium flame detector in a similar procedure using acetone instead of hexane and extracting for only one hour. In our experience, this gives lower recoveries at low Zinophos concentrations, possibly due in part to the large quantity of extraneous material in the extract or to the shorter extraction time.

TABLE I

Immediate Recovery of Zinophos From Soil

Sample Weight grams	Spike Rate p.p.m.	Recovery percent
Felida Silt Loam		
50	0.50	82
80	0.34	99
80	0.32	79
80	0.64	82
Lauren Sand Loam		
50	0.50	88
80	0.64	88
80	0.64	86
80	0.64	81
80	1.61	80
Burntbridge Silt Loam		
50	0.50	94
50	0.50	88
Caldwell Silt Loam		
50	0.50	85
80	0.64	97
80	0.64	90
80	0.64	79
20	0.85	76
19	0.85	96
22	0.85	79 Ave. 82%
18	0.85	93
19	0.85	75
22	0.85	76

TABLE I (cont.)

Sample Weight grams	Spike Rate p.p.m.	Recovery percent
Caldwell Silt Loam (cont.)		
20	4.2	91
20	4.2	100
21	4.2	96 Ave. 97%
22	4.2	98
20	4.2	97
17	4.2	100
21	4.1	77
20	4.1	94
20	4.1	88 Ave. 92%
20	4.1	95
21	4.1	100
21	4.1	103

TABLE II

Recovery of Zinophos From Caldwell Silt Loam
After Storage at -5° to -10° C.

Time of Storage days	Spike Rate p.p.m.	Sample Size grams	Recovery percent
49	5.06	40	92
49	5.06	58	93
114	5.06	50	78
191	3.47	99	78

Acknowledgments

The author gratefully acknowledges the technical assistance of John Neuman and the advice and encouragement of Richard C. Maxwell in this work.   References

(1) Coahran, D. R., Bull. of Environmental Contamination and Toxicology, In Press.

(2) Getzin, L. W., J. Econ. Ent. 59, 512. (1966)

(3) Kiigemagi, Ulo and Terriere, L. C., J. Ag. & Food Chem. 11, 293. (1963)

DDT Contamination of Feed Grains and Forages in Pennsylvania[1]

by H. Cole, D. Barry, and D. E. H. Frear
Pesticide Research Laboratories
Departments of Plant Pathology and Entomology
Pennsylvania State University, University Park, Pennsylvania

Introduction

It is well known that very small quantities of DDT may be found in meat, milk and poultry products (1, 2). These range from the lowest limits of confidence of the analytical method upward with an occasional instance of residues of 1 p.p.m. or higher. It is also recognized that DDT contained in and ingested with feed may accumulate in animal or poultry body tissues or be excreted in eggs or milk.

Federal tolerances established by the U. S. Food and Drug Administration for DDT in raw animal and poultry products range from 0 to 7 p.p.m. depending on the product. For meats such as beef which have DDT tolerances of 7 p.p.m., the tolerance can be met quite easily with routine care in selection of the feeds and the elimination or reduction in the use of DDT on the farm. With other products such as eggs and milk where zero tolerances are in effect it often has been difficult or impossible for farmers to produce products free from DDT in spite of all efforts to eliminate recognized sources of DDT exposure.

[1]Authorized for publication on August 15, 1966 as paper No. 3171 in the Journal Series of the Agricultural Experiment Station.

212

In an attempt to learn of the source of this DDT contamination, samples of forages and feeds to be fed to dairy animals, beef cattle, and poultry including broilers and laying hens were collected from Pennsylvania farms.

Methods

One hundred eighty-six (186) samples from 82 farms were collected during the calendar year 1965.

Samples were stored in moisture-proof polyethylene bags. All samples contained approximately 5 pounds of material collected from 20 different sites in the lot, thus a hay sample represented a composite from 20 bales and a grain sample a composite from 20 bags or 20 sites in a bin. Bacause of the impossibility of top-to-bottom sampling of silos each silage sample represented a composite of 20 sites from the level exposed at the time of sampling. Samples of silage were kept frozen until they were extracted.

In all instances these samples were from feed or forage grown on the farm or purchased from commercial feed distribution firms and were intended for livestock or poultry on the farm.

Samples were collected only from those farmers who kept accurate records of pesticide applications and could answer questions on the history of pesticide treatments on their farms. Thus it is recognized that the samples analyzed in this study can not be considered a truly random selection of feed stuffs in the statistical sense because there was a built-in bias in selecting farms for the collection of the feed samples.

DDT had not been used on any of the farms where crops were
sampled for at least three years prior to the collection of
samples.

The samples were classified as follows: pure alfalfa hay,
mixed legume and grass hay, alfalfa-grass silage, corn silage,
farm-grown grains, corn purchased from feed mills or distribution
centers, commercial mixed feeds, and commercial feed supplements
and concentrates. The latter three classes of feed were not produced on the farms, and hence were of unknown history. All forages sampled were grown on the farm where the sample was taken.

Analytical Methods

The samples of dry hay and whole grains were ground in a
Wiley Mill. The mixed feeds and concentrates were not re-ground.
A 100-gram sample of each of the dried materials was extracted for
16 hours in a large Soxhlet apparatus with redistilled n-hexane.
The extract was concentrated in a Kuderna-Danish apparatus to a
volume of two ml. Samples of silage were chopped in a Hobart food
chopper, and a 100-gram aliquot blended in an Omni-Mixer for several
minutes with a mixture of redistilled n-hexane and isopropyl alcohol
(2:1). After the solvent was decanted off, the sample was blended
again with chloroform-methanol (1:1). The extract was combined
with the n-hexane-isopropyl alcohol extract, washed with water in
a separatory funnel to remove the alcohols, filtered through anhydrous sodium sulfate to remove most of the water, stored over
anhydrous sodium sulfate, and later concentrated in a Kuderna-Danish

evaporator to two ml. This double extraction of fresh plant materials has been found to be more efficient than extraction with one solvent mixture (3).

Aliquots of the concentrated extracts were injected into a Research Specialties gas chromatograph equipped with five-foot glass columns packed with Chromosorb W using DC 200 and QF-1 as stationary phases; these instruments were equipped with electron capture detectors. The limit of confidence of this analytical procedure was 0.003 p.p.m. for p,p'-DDT. Samples containing less than this amount were considered to have no DDT residue.

Results and Discussion

Table 1 summarizes the information obtained from analysis of the collected samples. The results given are for p,p'-DDT. Although other pesticide materials as well as other isomers of DDT and its metabolites were occasionally present the occurrences of these other compounds were irregular and it was not considered advisable to include them in this report. The results are recorded in p.p.m. based on the feedstuff as sampled. No attempts were made to determine moisture contents or to convert results to standard moisture levels. It should be recognized in evaluating the results that the moisture contents of the two silage categories were higher than those of the hay and grain samples.

On this basis 54 percent of the forage and 62 percent of the grain and supplement samples contained detectable quantities of DDT ranging from 0.003 p.p.m. to a high of 0.150 p.p.m. for forage and

TABLE 1

p,p'-DDT Occurring in Feed Stuffs Sampled

| Forages | Total Number Samples | No Residue | Number of samples in p.p.m. ranges listed below | | | | Maximum level detected |
			0.003-0.010	0.011-0.050	0.051-0.100	>0.100	
Alfalfa hay	50	20	4	21	4	1	0.14
Mixed hay	24	17	0	6	0	1	0.14
Alfalfa grass silage	12	5	3	3	1	0	0.06
Corn silage	18	6	3	7	1	1	0.15
Total	104	48	10	37	6	3	0.15*
Feed grains and supplements							
Grains grown on the farm[1]	26	10	3	10	2	1	0.33
Purchased grain[2]	12	5	0	1	5	1	0.14
Commercial mixed feed	34	13	2	15	2	2	0.19
Commercial feed supplements	10	3	1	3	2	1	0.12
Total	82	31	6	29	11	5	0.33*

[1] Oats, corn, barley.

[2] All samples were corn.

* Maxima.

0.330 p.p.m. for grains. In both categories the majority of the samples that contained DDT ranged from 0.011-0.050 p.p.m. with 66 percent of the forages and 57 percent of the feeds falling within that range.

In light of these levels found in forages and feeds it is to be expected that small quantities of DDT will be present in meat, milk and poultry products from animals and poultry consuming these feeds.

The source of the p,p'-DDT in these forages and feeds is more difficult to determine. Contamination of other pesticides with DDT at the time of manufacture is possible but it seems unlikely that this form of contamination would be widespread. It has been pointed out that at least one chlorinated hydrocarbon insecticide is absorbed into the plants through the roots (4); this probably occurs on soils containing DDT. Another possibility is aerial drift of DDT from other agricultural operations in the vicinity where DDT may be used on potato fields or orchards, for example. This would account for some erratic high levels observed from time to time but again would not be expected to be a widespread source of contamination.

Some recent results by Antommaria et al. (5) working with DDT in airborne particulates may offer a clue to a major source of DDT contamination of feeds. In this June to December study these workers found that particulates in the air over Pittsburgh contained a maximum of 1.14 μg (on respirable particulates) and 1.22 μg

217

(on non-respirable particulates) of DDT per 1000 cubic meters of air sampled. The highest levels occurred in the July 6-20 collection period. Although these quantities are extremely minute, if particulates in air in other portions of the state contain DDT and if some of these particulates eventually land on soils and crops this may explain the occurrence of DDT in feeds grown at considerable distance from any known DDT applications.

Literature Cited

1. Durham, W. F., J. F. Armstrong and G. E. Quinby. Arch. Environ. Health 11, 641-647 (1965).

2. Ramsey, D. C. Assoc. South Agr. Workers Proc. 62, 118 (1965).

3. Mumma, R. O., W. B. Wheeler, D. E. H. Frear and R. H. Hamilton. Science, 152, 530-531 (1966).

4. Wheeler, W. B., D. E. H. Frear, R. O. Mumma, R. H. Hamilton and R. C. Cotner. In press.

5. Antommaria, P., M. Corn and L. DeMaio. Science 150, 1476-1477 (1965).

Bulletin

Contents

Bulletin of Environmental Contamination and Toxicology

AIMS AND SCOPE

The Bulletin of Environmental Contamination and Toxicology will provide rapid publication of significant advances and discoveries in the fields of pesticide residue research, air, soil, and water contamination and pollution, methodology, and other disciplines concerned with the introduction, presence, and effects of toxicants in the total environment.

Results of current research will be presented as brief reports providing information which is potentially useful to all individuals concerned with environmental contamination.

The articles will be free from restrictions imposed by purely scientific journals, particularly with respect to completeness of the studies reported and the attendant delays in publication.

Descriptions of new methods, procedures, or techniques shall be sufficiently detailed so as to permit direct application in other laboratories.

Review articles and obvious abstracts of papers forthcoming in other publications are not invited and probably will not be acceptable:

Articles suitable for inclusion shall be relatively short (less than 2,000 words) and will be prepared following specific instructions to permit reproduction by the photo-offset process from the original manuscript.

It is the hope of the Editorial Board that this Bulletin will provide a meeting ground for researchers who daily encounter problems related to the contamination of our environment and who welcome opportunities to share in new discoveries as they occur.

The Bulletin will be issued six times a year. This will be raised to 12 issues annually as demand increases.

Published bi-monthly by SPRINGER-VERLAG NEW YORK INC., 175 Fifth Avenue, New York N. Y. 10010, telephone (212) 673-2660. *Six issues per year. Subscription price: $15 per year for institutions. Special rates available for individual subscriptions for personal use only.* All orders must be accompanied by payment.

Effect of Surfactants on Soil Bacteria

by Ludwig Hartmann
Head, Laboratory of Engineering-Biology
Technische Hochschule Karlsruhe, Germany

Waste water reclamation is becoming the method of choice for solving water problems in arid climates. An example of this procedure is the Whittier Narrows Project (1), in Southern California, which was established to study all the technical, biological and biochemical problems connected with recharging sewage plant effluent into the groundwater by way of infiltration beds.

In Central Europe, the agricultural use of raw and treated sewage results in similar problems. Among these are the effects of residues of synthetic substances, for example surfactants, on agricultural crops and also on the organisms living in the soil. Recharge of sewage into the groundwater by means of infiltration beds and the agricultural use of sewage in crop production can only be practiced if the top soil and its biota are relatively undisturbed.

Studies on the influence of surfactants on plants (2) and bioadsorption phenomena associated with newer detergents (3) have already been reported.

The results reported in this paper are from experiments designed to reveal whether or not, and to what degree, the natural bacterial flora of soils is affected by surfactants.

219

Bulletin of Environmental Contamination & Toxicology,
Vol. 1, No. 6, 1966, published by Springer-Verlag New York Inc.

Methodology

The experiments were performed using the replica plating method of Lederberg and Lederberg (4,5). A dilution of the soil material to be tested was used to inoculate a Petri-dish containing 10% Trytic Soy Broth and 2% agar (pH: 7.0). After colonies had developed (48 hrs. at 20 C) they were transferred by the Lederberg method to other Petri plates containing the same medium augmented with the following concentrations of detergents: 0, 10, 20, 40, 75 and 100 ppm. After two days of incubation, the growth on the new plates was compared with that of the original culture and the missing colonies counted.

The soils used were from an oak forest and from a grassy field. Both soils belong to the para-brown-soil-type. In addition, two samples of surface water were tested.

Results and Discussion

The results summarized in the Table indicate that soil bacteria are more sensitive to surfactants than water-bacteria. In almost every case the losses of bacterial colonies from the soil samples are considerably higher then those of the water samples. This may be due to the fact that a greater part of the soil bacteria belong to the gram-positive type, while water-borne bacteria are in most cases gram-negative. The higher sensitivity of gram-positive bacteria to anionactive detergents has been reported previously (5).

The second striking result is that with increasing soil

Table. Effects of Surfactants on Bacteria

		Percentage Loss of Bacterial Colonies								
		with Depth (cm.) in							in	
Type	conc.	forest soil				grass field			water	
	ppm	0	4	8	15	0	2	4	brook	well
Alkylbenzenesulfonate										
(old, hard type)	10	10	0	55	20	0	3	25	0	0
M.W. 348	20	25	38	67	39	10	6	25	0	0
	40	55	81	67	53	14	12	42	0	0
	75	60	94	72	72	17	15	42	0	0
	100	60	94	72	75	17	18	58	0	0
Alkylbenzenesulfonate										
(new, soft type)	10	18	10	12	29	0	3	9	0	0
M.W. 348	20	35	30	12	47	14	3	9	0	19
	40	51	65	12	80	14	9	18	0	19
	75	51	75	24	83	22	9	37	3	19
	100	51	75	41	85	22	25	37	3	19
Ethersulfate										
M.W. 439	10	0	0	8	11	0	0	0	0	5
	20	0	0	15	21	7	3	5	0	5
	40	4	0	23	37	7	3	5	0	5
	75	4	11	23	40	7	3	5	5	5
	100	4	17	23	42	7	3	5	5	5
Alkylacrylsulfonate										
	10	0	8	54	45	4	4	0	0	0
	20	0	23	54	48	7	7	4	0	0
	40	10	39	54	68	11	7	8	0	0
	75	16	46	85	90	18	18	17	0	0
	100	27	62	85	90	18	18	25	0	0
Oil-acid hexa-										
methyleneimid	10	23	28	55	90	4	4	0	0	0
	20	46	46	82	100	8	10	0	0	0
	40	51	82	82	100	8	10	18	0	0
	75	51	82	82	100	12	14	18	0	0
	100	51	82	82	100	12	18	23	0	0
Oxethylized										
octylphenol	10	2	25	9	39	0	0	4	4	0
	20	50	69	73	87	0	4	17	11	6
	40	58	87	91	96	0	4	25	11	6
	75	58	87	91	96	6	4	29	11	6
	100	58	87	100	100	6	8	29	22	6
Fatty alcohol sulfate										
M.W. 334	10	4	0	9	0	0	-	0	0	0
	20	4	0	18	4	0	-	0	0	0
	40	7	8	27	43	0	-	0	0	7
	75	11	15	27	67	0	-	5	0	7
	100	14	23	46	67	0	-	11	11	7

depths the percentage of surfactant-sensitive bacteria also increases, while the biocoenosis of the soil surface is in most cases only slightly affected. This can be seen, for example, in the experiment with forest soil and the detergent "oil-acid-hexamethyleneimide". In this case the loss of bacterial colonies by adding 10 ppm of surfactant to the surface sample is only 22 percent. However, 90 percent of the colonies are lost in the sample taken at 15 centimeters depth. With higher concentrations of surfactants, the loss is even more evident. The application of 100 ppm of the surfactant reduces the number of bacterial colonies in the surface sample by 51 percent. In the 15 centimeter sample, no colonies developed.

There are not only differences between soil and water samples but also differences between different soil samples. With forest soil the previously mentioned substance, "oil-acid-hexamethyleneimide," shows the greatest toxicity, but in the soil samples from the grassy field, this toxicity is surpassed by the two alkyl-benzene-sulfonate-types.

It is also interesting to compare the new so-called "soft" alkylbenzenesulfonate with the old so-called "hard" type. The toxicity of the new type seems to be slightly higher than the old for both water and soil borne bacteria.

These results do not permit conclusions on the biochemical mechanism of the observed toxicities. Since the detergents are rather large molecules, they probably do not enter the living bacterial cell. By their very nature, these substances tend to

accumulate at the borderline between two different phases, and it
is probable, that the observed toxicity is due to reactions at the
cell surface. Depolarization of the cell membrane by the absorption of detergents could result in decreased absorption of
essential nutrients and decreased release of toxic metabolic
products resulting ultimately in the death of the organism. This
was also indicated in previously published results (6).

The adsorption of ABS leads to a reduction of the food
uptake and oxygen consumption of aerobic bacteria. This occurs
with both gram-positive and gram-negative bacteria. The
gram-negative bacteria absorb detergent at a slower rate and,
since this phenomenon is dependent on the free and adsorbed
phases being in equilibrium, higher concentrations of ABS must be
used for gram-negative organisms.

These results show that the presence of detergents in
sewage-derived irrigation water is apt to destroy the natural
composition of the bacterial population in the soil. Certain
sensitive species will vanish, while others may be stimulated,
but the original biological equilibrium will be disturbed to a
degree which will have adverse effects on the purification
capability of the top soil.

In natural soils with high concentrations of substances
capable of adsorbing detergents this danger will be reduced, but
will not be totally avoided. Experiments in progress have shown
that while the carbon dioxide content in the atmosphere of
undisturbed soil usually amounts to 3 to 4 percent, the

CO_2-content in soil columns (15 cm high) irrigated with soft type ABS enriched tap water dropped to values as low as 0.5 percent. These results indicate that even the "soft" - partly degradable types of detergents cause a disturbance in the soil biota.

Acknowledgment

This work was supported in part by a grant from the Wirtschaftsministerium Baden-Wurttemberg.

References

1. J. E. MCKEE and F. C. MCMICHAEL, Research on waste water reclamation at Whittier Narrows, State Water Quality Control Board, Sacramento, Calif. 1963.

2. L. HARTMANN, Gas-Wasserfach. 107, 251 (1966).

3. L. HARTMANN and H. MOSEBACH, Journal Tenside (in press).

4. J. E. LEDERBERG and E. M. LEDERBERG, J. Bacteriology 63, 399 (1952).

5. L. HARTMANN, Gas-Wasserfach (in press).

6. L. HARTMANN, Biotechnology and Bioengineering 5, 331 (1963).

Alkyl Benzene Sulfonate
Effects on Stream Algae Communities

by Charles E. Hicks
Utah State Department of Fish and Game, Salt Lake City, Utah

John M. Neuhold
Utah State University, Logan, Utah

ABS is toxic to fish and other aquatic organisms. It also is toxic at varying concentrations to lower life forms at all trophic levels within the aquatic environment. Algae represents the level of the primary producer. Those forms ranking above it are dependent upon it for their energy supply. A reduction in the productive ability of algal communities will result in a reduction of the total energy flow and a consequent reduction in the productive capacity of the entire community. The object of this investigation was to determine the effects of alkyl benzene sodium sulfonate on the productivity of stream algal communities.

The major types of synthetic detergents have diverse effects on a wide variety of organisms. Manganelli and Crosley (1) found that synthetic detergents appear to denature the cytoplasmic membrane and damage the cytoplasm of the protozoa and higher organisms in activated sludge. Tomcsik (2) indicated that surface active agents denature the cytoplasmic membrane of Bacillus sp. and damage protoplasts.

Work on plants in relation to detergent effects is limited. Degence (3) observed no adverse effect when four water plants were grown in concentrations of 40 mg/l of anionic detergent. Jenkins, Klein, and McGauhey (4) found that ABS was concentrated in barley and sunflower roots when grown in water cultures. At 10 mg/l ABS the sunflowers were retarded 20 percent, and at 50 mg/l the plants were retarded 65 percent. Other sunflower plants in water cultures containing 10 mg/l ABS showed distinct signs of chlorosis. Lupinus albus, which was grown in fertilized soil with up to 20 mg/l ABS, was unaffected.

Hynes (5) showed that Cladophora domerate placed in shallow dishes containing 50 mg/l ABS suffered chloroplast destruction and chlorophyll loss into the water. Algae which had been placed in 25, 20, and 10 mg/l ABS died after three weeks. That which was placed in 5 mg/l ABS appeared unhealthy. In 2.5 mg/l and in the controls, the plants appeared unaffected. Both Ranunculus pseudofluitan and Potomogeton pectinatus were seriously affected by 2.5 mg/l ABS.

Experimental

Two types of algal communities were used in experiments. Vaucheria communities were obtained from the Logan River, a swift moving stream. Cladophora communities were obtained from a standing water aquarium. Both types of communities are found attached to rocks in relatively large mats. Parts of the algal mats were transferred to experimental containers. These samples were considered as the experimental unit. No attempt was made to obtain pure cultures. Attendent invertebrates and bacteria were considered part of the community.

The ABS used in all experiments was of the same type. The trade name for the compound is Nacconol NRSF. The compound is a propylene derivative and is 92.5 percent active ABS. A stock solution of 5,000 mg/l ABS was made for each experiment. Aliquots of the stock solution were added to the experimental containers to obtain desired concentrations.

The parameter measured was that of the rate of C^{14} uptake by the communities. The primary production in an aquatic ecosystem is dependent upon the rates at which photosynthetic organisms fix inorganic carbon and produce organic material. The use of C^{14} presents a comparatively simple procedure, and most important, is a very sensitive method for measuring production (6). Ryther (7) states that C^{14} uptake represents a measure of net production and not gross production. Strickland (6) says that C^{14} uptake experiments generally give a measure of photosynthesis that is somewhere between the net and gross values, possibly nearer the former. C^{14} has mostly been applied to work related to phytoplankton productivity. However, the same basic principles apply when used with benthic periphyton.

Conditions were varied by subjecting community samples to different concentrations of ABS for different periods of time. All experiments included completely randomized factorial arrangement of treatments. The basic response, counts per minute per milligram algae, was subjected to an analysis of variance to determine the effects of time and ABS concentrations.

All experiments included a 4 x 5 factorial arrangement of treatments. The experiments for both Vaucheria and Cladophora communities in high and low ranges of ABS concentrations consisted of five one-gallon glass containers, twenty-five 60 ml. glass bottles and a water trough employed as a constant temperature bath. The algal communities were placed in the gallon containers. Each container held one liter of water. Appropriate concentrations of ABS were added. Algal communities were subsampled and the samples were placed in 60 ml. bottles at four time periods: 12 hours, 24 hours, 48 hours, and 96 hours. Four samples were taken from each of the respective gallon containers. This represented four replications of each treatment or concentration of ABS. C^{14} was added in 0.25 uc aliquots to each bottle. Five of the 60 ml. bottles, one representing every ABS concentration, were covered with aluminum foil to estimate error due to surface adsorption of C^{14} by the algal communities and account for loss due to respiration in the dark. The bottles were placed in the trough and flooded with 700 foot-candles of light.

The concentrations of ABS used in the Cladophora-Vaucheria experiments were 0, 5, 25, 50 and 100 mg/1. In both of these experiments, the water used was from a standing water aquarium. In all other laboratory experiments tap water was used after it was passed through an active carbon, sand, and gravel filter. The chemical quality of the tap water in mg/1 included 42.0 Ca^{++}, 18.0 Mg^{++}, 4.0 K^{+}, 0.5 NH_4^{+}, 1.0 Cl^{-}, 8.0 $SO_4^{=}$, 209.0 HCO_3^{-}, 0.44 $NO_3^{=}$, 1.5 $PO_4^{=}$, and a conductivity of 329 umhos. The aquarium water included 44.0 Ca^{++}, 85.0 Mg^{++}, 8.0 K^{+}, 0.3 NH$_4{}^{+}$, 10.0 Cl^{-}, 508.0 HCO$_3{}^{-}$, 44.0 NO$_3{}^{=}$, 95.0 PO$_4{}^{\equiv}$, and a conductivity of 924 umhos.

Concentrations of 0, 4, 8, 16 and 32 mg/l ABS were used in experiments involving "soft" and "hard" tap water. Vaucheria was used in both runs. "Soft" water contained 2 mg/l total hardness as .calcium carbonate. "Hard" water contained 180 mg/l total hardness as calcium carbonate.

In all experiments, the algal communities were killed with a 4:1 solution of concentrated hycrochloric and glacial acetic acids. The samples were flushed with distilled water, dried at 90° C for 48 hours and weighed. The samples were finally pulverized and distributed evenly on planchetts.

All samples were counted for five minutes in a 2 pi proportional counter. The resulting response data was expressed as counts per minute per milligram algae.

Results and Discussion

The analysis of variance for the Cladophora experiment showed that ABS concentration, time, and their interaction had an effect on the uptake of C-14 by the community (Table 1). Both treatments have a negative influence on the assimilation of C^{14}. If the main effect of concentration is considered alone, 5 mg/l stimulates assimilation of carbon. The time factor contributed to the reduction of the productivity of the algal community.

The results of the Vaucheria experiment were slightly different: Only ABS concentration had an effect on C^{14} assimilation. Five mg/l ABS seemed to stimulate uptake.

TABLE 1

Treatment means and analysis of variance of the amount of C^{14} (counts per minute per mg) assimilated by <u>Cladophora</u> and <u>Vaucheria</u> in aquarium water treated with alkyl benzene sulfonate.

ABS Conc (ppm)	Species	Time of Exposure				Mean
		12 hrs.	24 hrs.	48 hrs.	96 hrs.	
0	Clado.	277.48	314.27	147.23	56.41	198.87
	Vauch.	65.28	79.45	71.06	107.81	80.90
5	Clado.	381.23	326.04	184.77	69.31	240.33
	Vauch.	56.82	93.40	96.22	103.75	87.55
25	Clado.	178.10	149.33	150.77	58.64	134.21
	Vauch.	16.10	12.85	6.30	5.51	10.19
50	Clado.	77.64	120.88	30.63	14.94	61.02
	Vauch.	15.25	10.85	2.96	2.77	7.96
100	Clado.	18.31	14.37	7.38	11.45	12.87
	Vauch.	6.39	3.76	1.85	2.43	3.61
Mean	Clado.	233.19	231.22	130.21	52.68	
	Vauch.	39.96	50.08	44.60	55.57	

Analysis of Variance

Source of Variation	D.F.	Cladophora		Vaucheria	
		Sum of Squares	Mean Squares	Sum of Squares	Mean Squares
Replications	3	161.19		574.20	
Concentration	4	563870.59	140967.63*	114456.90	28614.23*
Time	3	290079.22	96693.07*	1765.25	588.42
Time x conc.	12	178282.49	14856.87*	8569.69	714.14
Experimental Error	57	134451.01	2358.78	24353.36	427.25
Total		1166844.50		149719.40	

*Significance at the 0.5 percent level.

It is interesting to note that maximum uptake for <u>Cladophora</u> was higher than maximum uptake for <u>Vaucheria</u>. Both experiments indicated that as concentration increased, assimilation of the C^{14} decreased. In the run utilizing <u>Vaucheria</u>, the effect of concentration over time did not change significantly.

The point estimating equation determined from the coefficients accounting for 80 percent of the variation was:

$$Y = 350.5 - 5.78X_1 + 0.02X_1^2 - 3.03 \; X_2 + 0.003 \; X_1X_2,$$

where Y is C-14 uptake in counts/minute/mg of community, X_1 is
concentration of ABS in mg/l and X_2 is time of exposure in hours.
A surface representation of the interactions involved indicate a
linear drop in C^{14} uptake from a high point at 0 concentration and
the first time period (Fig. 1). The surface displayed a low response
at 75 mg/l and 96 hours.

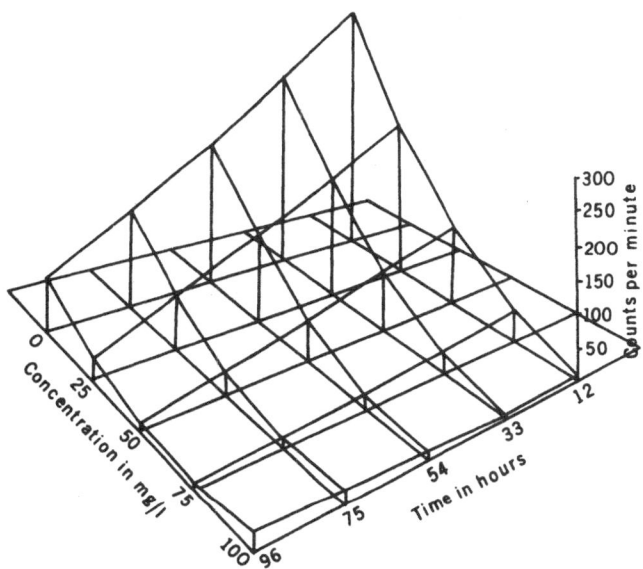

Figure 1. C^{14} uptake in counts per minute per mg of <u>Cladophora</u>
community as related to time of exposure to ABS
concentration.

The soft water experiment with ABS concentrations of 0, 4, 8,
16 and 32 mg/l indicated significance due to ABS concentration, time,
and their interaction. Response differences due to time and the
concentration x time interaction were found at the 0.5 percent level
of confidence. The differences due to concentration were found at
the 10 percent level (Table 2).

TABLE 2

Treatment means and analysis of variance of the amounts of C^{14} assimilation by Vaucheria in "soft" and "hard" water.

ABS Conc.(ppm)	Water	Time of Exposure 12 hrs.	24 hrs.	48 hrs.	96 hrs.	Mean
0	Soft	102.47	56.91	29.57	68.48	64.35
	Hard	186.78	55.21	18.49	20.38	70.20
4	Soft	66.94	81.20	38.90	36.51	55.88
	Hard	91.33	42.04	22.64	14.65	42.66
8	Soft	44.18	41.94	52.82	61.98	50.09
	Hard	106.06	24.46	32.44	21.35	47.07
16	Soft	25.93	85.63	41.34	47.74	50.16
	Hard	87.15	28.83	14.48	6.81	34.31
32	Soft	51.68	58.01	33.43	12.84	38.99
	Hard	21.05	14.64	15.14	3.89	13.68
Mean	Soft	72.80	80.92	48.88	56.88	
	Hard	98.67	33.63	20.63	13.41	

Analysis of Variance

Source of Variation	D.F.	Soft Water Sum of Squares	Mean Squares	Hard Water Sum of Squares	Mean Squares
Replications	3	2329.43		5583.44	
Concentration	4	5504.58	1376.14***	26768.50	6692.12*
Time	3	8191.85	2730.61**	90957.22	30319.07*
Time x Conc.	12	22093.46	1841.12*	35147.04	2928.92*
Experimental Error	57	34370.45	609.30	36542.94	641.10
Total	79	72849.77		194999.14	

*Significance at the 0.5 percent level.
**Significance at the 1.0 percent level.
***Significance at the 10.0 percent level.

The analysis of variance for the "hard" water experiment indicated significance due to differences in concentration, time, and their interaction (Table 2). There was an apparent stimulation of the uptake of C^{14} at low levels of ABS. Maximum counts in this experiment were comparable to those in the previous experiment with Vaucheria communities. The point estimating equation accounting for 58 percent of the variation was found to be:

$$Y = 116.79 - 3.99X_1 + 0.032X_1^2 - 4.09X_2 + 0.026X_2^2 + 0.032X_1X_2 \text{ (Fig. 2)}.$$

The decrease due to the main effect of ABS concentration exhibited
a linear relationship. The decrease due to the main effect of time
showed a quadratic relationship, inferring a slight recovery in
response over time. The interaction in this experiment included both
linear and quadratic terms. However, there was a general decrease
from the high point at 0 concentration and 12 hours to a low point
at an ABS concentration of 32 mg/1 and 54 hours.

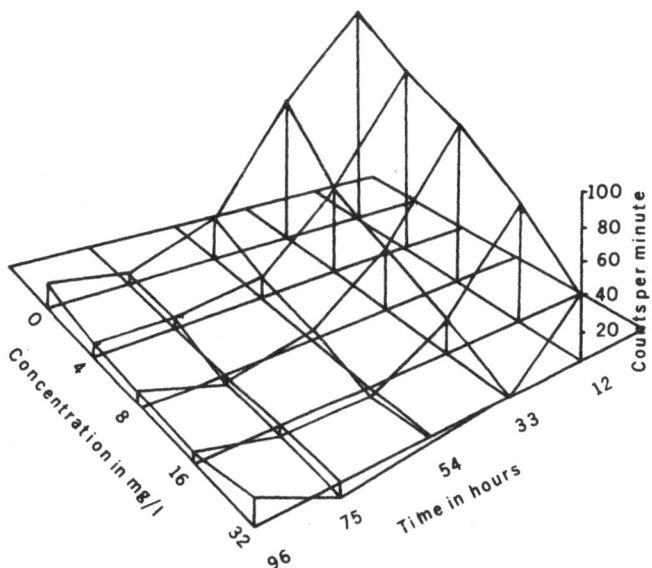

Figure 2. C^{14} uptake in counts per minute per mg of <u>Vaucheria</u>
community in hard water as related to hours exposed to
ABS concentrations.

The soft water experiment also displayed significant concentration,
time and interaction effects (Table 2). The response surface was not
easily defined. The best fit with linear and quadratic terms
accounted for only 37 percent of the variance. An empirical plot of

the surface, however, still is sufficiently revealing to merit description (Fig. 3).

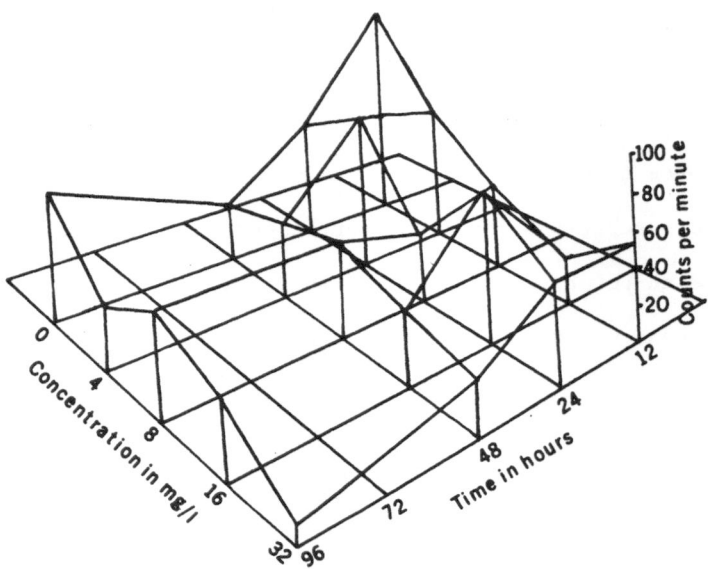

Figure 3. The empirical response surface of C^{14} uptake (counts per minute per mg) of <u>Vaucheria</u> communities in soft water treated with ABS concentration.

Upon examination of the tables of means and the figures, three effects become apparent: (1) ABS has a negative effect on C^{14} uptake for both algae communities, (2) the communities appear to partly recover their ability to assimilate C^{14} at extended exposures to high concentrations, and (3) a slight stimulation of C^{14} uptake appears to occur at abbreviated exposures to low concentrations.

A reduction in productivity is alluded to by the decrease in C^{14} assimilation. Ample evidence exists that the photosynthetic mechanism is affected by detergents, e.g., detergents fraquent chloroplasts (8) as well as make the chlorophyll-protein complex water soluble (9). Chlorophyll leaching from the cells was observed during the course of the experiment.

The apparent increase in C^{14} assimilation at extended exposures to high concentrations may be a function of the heterogenity of the community. Mortality of the algae may provide the basis for proliferation of reducer organisms which are less sensitive to the ABS concentrations. Such organisms as Sphaerotilus natans which exist on organic material are present in the community.

The apparent stimulation at low concentrations is a phenomenon which also has its parallels. Robinowitch (10) points out that most inhibitors appear to have a stimulating effect at low concentrations. Manganelli and Crosley (1) noted stimulated "slime" growth in low concentrations of detergent.

The differences that existed between Cladophora and Vaucheria communities (Table 1) illustrate species or adaptation differences. The endemic environment of both communities was changed for the experiment. Vaucheria was taken from a flowing water environment with a water quality similar to the tap water and placed into aquarium water without a velocity. The Cladophora community was subjected to a temperature change from 27° C. to 10° C.

Acknowledgement

The research reported here was conducted with the aid of NIH RG EF 00044.

235

Literature Cited

1. MANGANELLI, R., AND E.S. CROSLEY. Sew. and Ind. Waste 25(1):262-276. 1953.

2. TOMCSIK, T. Proc. Soc. Expt. Biol. Med. 89(3):459-463. 1955.

3. DEGENCE, VANDER ZEE ET ALIER. Journal Institute of Sewage Purification. Part 1. 1950.

4. JENKINS, D., S.A. KLEIN, AND P.H. MCGAUHEY. Water Pollution Control Federation, May: 636-654. 1963.

5. HYNES, N.B.N., AND F.W. ROBERTS. Ann. Appl. Biol. 50:799. 1962.

6. STRICKLAND, J.D.H. Fisheries Research Board of Canada Bull. 122, 172 pp. 1960.

7. RHYTHER, J.H. Deep Sea Research 2(2):134. 1954.

8. GIESE, A.C. Cell Physiology. 1962. W.B. Saunders Co., Philadelphia, 592 pp.

9. SMITH, G.M. The fresh-water algae of the United States, 1933. McGraw-Hill, New York, 716 pp.

10. RABINOWITCH, E.T. Photosynthesis. 1945. Interscience, New York, New York, 599 pp.

Forced Volatilization Cleanup for Gas Chromatographic Assay of Pesticide Residues[*]

by F. A. Gunther, R. C. Blinn,[**] and D. E. Ott
Department of Entomology
University of California Citrus Research Center
and Agricultural Experiment Station, Riverside, California

A major deterrent to the widespread routine application of

gas chromatographic segregative techniques and selective detectors

to qualitative and quantitative pesticide and other residue evalu-

ation has been "poisoning" of volatilization chamber and column by

extraneous extractives. This present note is prompted by the re-

cent publication by STORHERR and WATTS (1) of their "sweep co-

distillation" cleanup method, which is a version of our "forced

volatilization" cleanup technique [OTT and GUNTHER (2)] as ap-

plied to butterfat. The final versions of our apparatus were

best suited to cleanup of certain pesticides in butterfat, but

the intermediate versions, described herein and in part by

GUNTHER (3) were applicable with good recoveries to a variety

of substrate extractives including the very intractable extrac-

tives mixture from alfalfa. All these devices physically sepa-

rate gas chromatographable compounds from those compounds not

---

[*] Presented in essential part at the 139th Annual Meeting of the American Chemical Society, Division of Agricultural and Food Chemistry, St. Louis, Mo., March 1961.

[**] Present address: American Cyanamid Co., Princeton, N. J.

237

gas chromatographable and provide a simple and versatile approach to the cleanup of stripping solutions to eliminate many extraneous extractives for almost any subsequent analytical operations.

Method and Apparatus

This cleanup procedure consists of hypodermically injecting a sample solution into a reproducibly heated chamber, sweeping the volatile materials through an aerosol filter and away from those that possess little or no volatility, then trapping the volatiles for further gas chromatographic or other scrutiny; the less volatile residuum is mechanically discarded from the apparatus. If desired, a section of gas chromatographic column packing can be placed in the apparatus to achieve any desired degree of retention of unwanted low-volatility compounds; appropriate choices of nature and of amount of column packing will often provide compounds sufficiently cleaned up for direct ultraviolet or infrared assay, for example. Also, the device can be adapted for either macro- or microscale separations.

In Figure 1 is shown one version of the basic design of this device.

Shown in more detail in Figure 2 is the volatilization portion of this simple device.

An even simpler version shown in Figure 3 accomplishes the same purpose. One advantage of this latter design is ease of removing the non-volatile residuum, which adheres to a replaceable wad of clean glass wool placed near the inlet end. The design of any volatilization cleanup device may be varied to suit supplies,

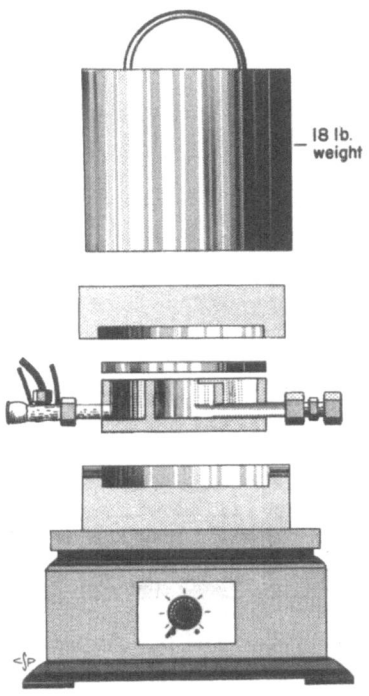

Fig. 1. Volatilization cleanup device constructed of aluminum. Component parts in ascending
order are: thermostatically controlled hot
plate, lower heat sink, volatilization chamber,
lid for volatilization chamber, (lid sealed
with a high-vacuum silicone lubricant), upper
heat sink, and 18-lb. steel weight. Reproduced
from GUNTHER (3) with permission

needs, and desires, but temperature profiles within the volati-

lization chambers must be reproducible.

In Figure 4 is shown the design of a special scrubber used

for trapping the volatile eluants with the necessarily high rate

of carrier gas flow, as discussed by OTT and GUNTHER (2). The

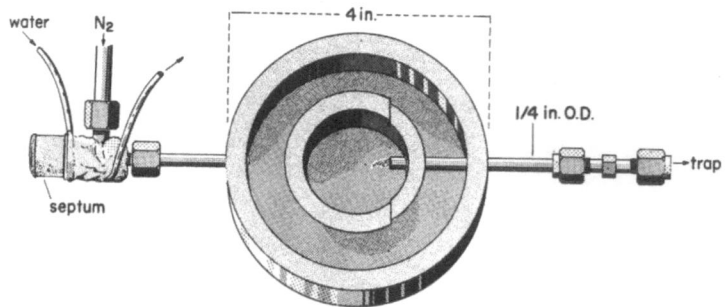

Fig. 2. Volatilization chamber of a cleanup device. The central chamber is usually filled with gas-chromatographic column packing on glass or quartz wool, and the outer section near the injection port is one-third filled with ignited sea sand. Reproduced from GUNTHER (3) with permission

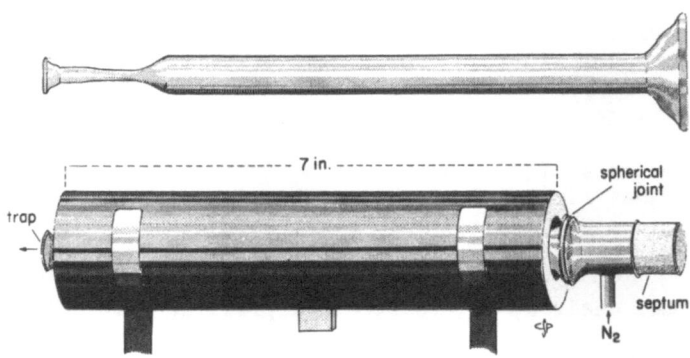

Fig. 3. Volatilization cleanup device constructed entirely of glass or quartz. Furnace is controlled by means of a variable resistor

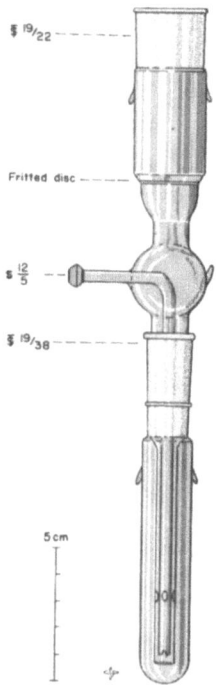

$\frac{19}{22}$

Fritted disc

$\frac{12}{5}$

$\frac{19}{38}$

5 cm

Fig. 4. Scrubber used for trapping volatiles in a high-flow gas stream (two-thirds actual size). The sintered glass disc is coarse porosity. The slot at the bubbler tip decreases bubble size. Emerging bubbles create a gas lift-pump, releasing slugs of solvent through the holes in the inner member and thus violently circulating the scrubbing solvent. At the end of a "run", the ⚓ joint is loosened and the liquid in the "funnel" drains down the inner member to rinse it. Reproduced from GUNTHER (3) with permission

tube is the conventional Kuderna-Danish type [GUNTHER and BLINN (4)], and the solvent of choice is placed both in the tube and in the fritted glass scrubber to prevent entrainment losses.

Presented in Table I are the recoveries achieved with several

TABLE I

Percent recoveries[a] of pesticides through volatilization cleanup devices

| | Construction material | | |
| | GLASS | | ALUMINUM |
Pesticide	Glass wool Packing	Glass wool & silicone-coated firebrick packing	Sea sand & silicone-coated firebrick packing
	(%)	(%)	(%)
Aldrin	94 ± 8	--	90 ± 7
Carbaryl[b]	50,62	64,67	66 ± 2
DDT	77 ± 2	82,77	80 ± 2 DDT (38% as DDE)
DDE[c]	103 ± 5	66,57	85 ± 5
p,p'-Dichloro-benzophenone[d]	90,75	77	93 ± 4
Dieldrin	100 ± 5	--	90 ± 4
Ethion	69,68	--	69 ± 1
Kelthane[e]	--	62,38	84 ± 3
Lindane[f]	--	--	85 ± 4
Malathion	44 ± 8	--	69 ± 7
Parathion	46,36	62,77	69 ± 7
Tetradifon	71 ± 7	51,52	80 ± 3

a/ By infrared assay except as indicated.
b/ Appeared as α-naphthol.
c/ Product from p,p'-DDT.
d/ Product from p,p'-DDT and Kelthane.
e/ Appeared as p,p'-dichlorobenzophenone.
f/ Percent recovery by microcoulometric titration after combustion.

purified pesticides to demonstrate broad versatility under empirical but uniform developmental conditions. The temperature was 250° C., the nitrogen flow rate was 250 ml./minute, and the collection period was 10 minutes.

Discussion

It must be emphasized that the above conditions were selected arbitrarily; optimum forced volatilization conditions for each compound of interest **must** be determined in advance of a residue study, then evaluated in the presence of the substrate extractives of interest. Work with actual stripping solutions, such as those involving DDT in alfalfa and in butter, indicates that/ inefficient _with agitation or "turnover" of extractives from an insufficient gas stream flow rate recoveries are affected by the amounts and natures of non-volatiles introduced into the apparatus. Preliminary de-waxing, selective adsorption, or partitioning may sometimes still be desirable to reduce the bulk of these "keepers." Also, effic-ient thin-layer distribution on the glass wool or sand plus more efficient heat transfer for volatilization can be improved; fine inert metal granules sometimes improve heat-transfer. A verti-cal unit achieved improved "sweeping" abilities out of masses of oily extractives [e.g., butter as discussed by OTT and GUNTHER (2)] by permitting a high carrier gas flow rate without carry-over of excessive amounts of non-volatiles.

It is the main purpose of this report to reaffirm the great promise of this completely physical type of cleanup, not only for gas chromatography where volatilization chamber and column "poisoning" are major deterrents to really reliable widespread use, but also for almost any residue evaluation in which cleanup is a problem. It must be emphasized that some pesticides will decompose thermally under these conditions (2,5,6 and cf. Table I) or may be attacked by the free alkali present even in (and on the surface of) borosilicate glass. Thus mg. quantities of p,p-DDT cycled several times through the glass version of Figure 3 were quantitatively dehydrochlorinated, and lindane yielded the usual mixed trichlorobenzenes; dehydrochlorination did not occur, however, when an all-quartz system was used.

References

1. R. W. STORHERR and R. R. WATTS, J. Assoc. Official Agr. Chemists 48, 1154 (1965).

2. D. E. OTT and F. A. GUNTHER, J. Agr. Food Chem. 12, 239 (1964).

3. F. A. GUNTHER, Adv. Pest Control Research 5, 191 (1962).

4. F. A. GUNTHER and R. C. BLINN, Analysis of Insecticides and Acaricides, pp. 231-233 (1955), Interscience, New York.

5. D. E. OTT and F. A. GUNTHER, Residue Reviews 10, 70 (1965).

6. F. A. GUNTHER, J. H. BARKLEY, R. C. BLINN, and D. E. OTT, S.R.I. Pesticide Research Bull. 2 (No. 2), 3 (1962).

Paper No. 1682, University of California Citrus Research Center and Agricultural Experiment Station, Riverside, Calif.

Forced Volatilization or Sweep Extraction of Organochlorine Pesticide from Vegetable Extracts

by R. Mestres and F. Barthes
Faculté de Pharmacie
Université de Montpellier, France

The sensitivity of the electron capture detector to numerous natural products requires a good cleanup of vegetable extracts when this type of detector is used for organochlorine pesticide residue analysis (1).

OTT and GUNTHER (2) achieved some partition by the forced volatilization with nitrogen of organochlorine pesticides from butterfat and alfalfa (3), followed by gas chromatographic analysis with the Dohrmann microcoulometric detector. More recently, STORHERR and WATTS (4) used an equivalent cleanup procedure to separate organophosphate pesticides from vegetable extracts; they analyzed the residues by gas chromatography with a sodium thermionic detector.

We proposed to get sufficient cleanup to analyze organochlorine pesticide residues by gas chromatography using an electron capture detector in a similar way (1).

VOLATILIZATION CLEANUP APPARATUS

The apparatus for cleanup (Fig. 1) is similar to that of GUNTHER et al. (3) and consists of an electric furnace for microanalysis with an alternostat selector of temperature. This

245

Bulletin of Environmental Contamination & Toxicology,
Vol. 1, No. 6, 1966, published by Springer-Verlag New York Inc.

cylindrical furnace a (22 cm. long, 1.8 cm. I.D.) admits a large

borosilicate tube b (23 cm. long, 1.6 cm. O.D.) that ends with a

male T joint. A Teflon sleeve joins it to·the female joint of the

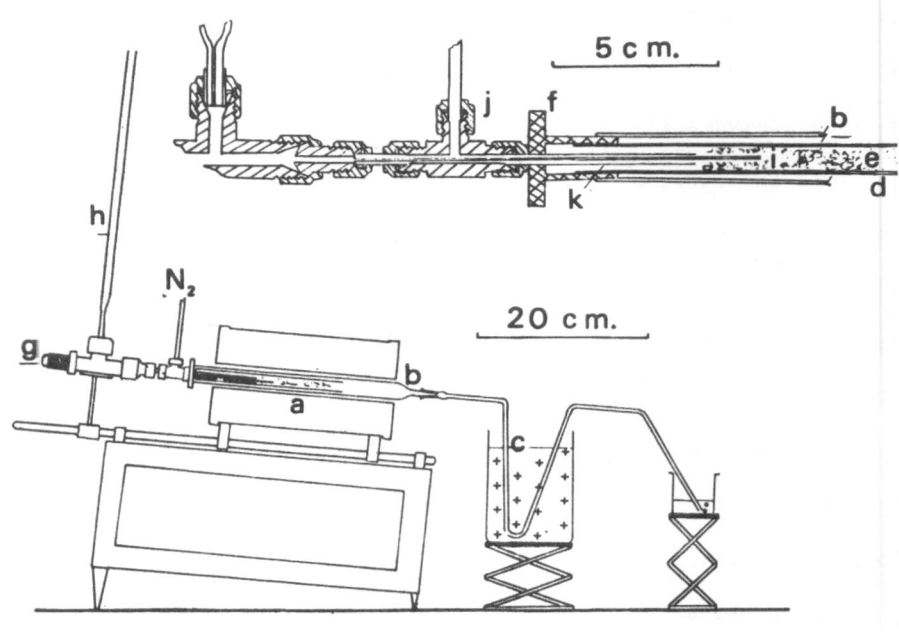

Fig. 1. Forced volatilization or sweep extraction apparatus.
See text for details.

bubbling tube c. This is bent as shown, is surrounded with ice in the large beaker, and terminates in a small beaker containing petroleum ether. A small borosilicate tube d (17 cm. long, 1.0 cm. O. D.) set inside the large tube b contains 0.2 g. of borosilicate wool e renewed for every cleanup.

A Teflon stopper f joins the large tube and the inside tube to the metal solvent and nitrogen entrance structure. This structure includes: a micrometer valve g (1/4") for regulation of the flow of petroleum ether. This valve receives a vertical glass tube h which serves as a solvent tank of 10 ml. capacity. A Teflon sleeve protects and seals the joint. The petroleum ether flows from the valve and reaches the borosilicate wool through a stainless steel pipe i (1.5 mm. O.D.). This pipe goes through a T-joint j (1/8") which allows entrance of the nitrogen. Nitrogen arrives at the borosilicate wool through a stainless steel pipe k (3.1 mm. O.D., or 1/8") that encases the solvent pipe i. This structure allows separate regulation of solvent flow and nitrogen flow.

METHOD

The vegetable sample is extracted, for instance by acetonitrile; the residue in the acetonitrile is then partitioned into petroleum ether (1), and the ether extract is concentrated 10:1.

With the apparatus assembled, regulate the furnace temperature at $260° \pm 5°$ C. Cool the bubbling tube with ice while the end dips into petroleum ether. Regulate the nitrogen flow to ca. 200 ml. per min. Then introduce into the vertical glass "tank" 1 ml.

of petroleum ether vegetable extract corresponding to 10 g. of sample. Flow it into the system, then flow through the vertical tube 20 ml. of petroleum ether, at a constant rate, over a period of one hour.

Rinse the bubbling tube with petroleum ether; collect the washings in the receiving beaker. Concentrate this purified extract by transfer to a graduated conic tube and reduction of volume to 1 ml.

Inject into the chromatograph 2 μl. of this purified extract, a quantity equivalent to 20 mg. of the parent sample.

RESULTS

In this way we have purified extracts of different fruits and vegetables such as apples, beans, cabbages, carrots, green peas, oranges, potatoes, radishes, and tomatoes.

The electron capture gas chromatograms of these extracts show that the purification is satisfactory, with no "background" interferences for the organochlorine pesticides. However, samples of carrots have contained two or three unidentified substances as revealed by the electron capture detector; these compounds act like organochlorine insecticides upon chromatography through Florisil columns.

The method, when used precisely as described above, gives good separation between pesticides and the normally extracted interfering compounds. We have fortified crude fruit and vegetable extracts with from 0.02 to 0.12 p.p.m. of several pesticides, with good recoveries, as shown in Table I.

TABLE 1

Recovery of pesticides added to uncontaminated extracts of fruits and vegetables (apples, green peas, potatoes, radishes, and tomatoes)

Pesticide	Recovery (%)	Pesticide	Recovery (%)
BHC	100	o, p'-DDT	100
Lindane	100	p, p'-DDT	95-100
Heptachlor	100	Dieldrin	100
Heptachlor epoxide	100	Endrin	100
Aldrin	100	Chlordan	100
o, p'-DDE	100	Toxaphene	95
p, p'-DDE	100	Kelthane	100
		Methoxychlor	75

CONCLUSIONS

Selective partition by vapor solvent sweeping, on borosilicate wool, of organochlorine pesticides from fruit and vegetable extracts in the present apparatus and with the present technique provides sufficient cleanup for subsequent electron capture detector analysis. Amounts of organochlorine pesticides from 0.02 to 0.1 p.p.m. in fruits and vegetables are quantitatively recovered and can be analyzed by this method. This study, like those of GUNTHER et al. (2,3) with organochlorine pesticide residues in butterfat and in alfalfa and of STORHERR and WATTS (4) with organophosphate pesticides in several fruits and vegetables confirms that forced volatilization cleanup is a promising method.

REFERENCES

(1) R. MESTRES and F. BARTHES, Trav. Soc. Pharm. Montpellier 26 in press (1966).

(2) D. E. OTT and F. A. GUNTHER, J. Agr. Food Chem. 12, 239 (1964).

(3) F. A. GUNTHER, R. C. BLINN, and D. E. OTT, 139th Meeting Amer. Chem. Soc., St. Louis, Mo. (March 1961); Bull. Environmental Contamination and Toxicol. 1, 237 (1966).

(4) R. W. STORHERR and R. R. WATTS, J.A.O.A.C. 48, 1154 (1965).

Insect Toxicological Studies of Maleic
Hydrazide Translocated in the Potato Plant

by W. N. YULE, I. HOFFMAN, and E. V. PARUPS
Research Branch, Canada Department of Agriculture, Ottawa

Doubts about long-range biological effects due
to the use of agricultural chemicals in the production
and storage of foodstuffs remains a cause for great
concern. Attempts have been made through legislation
to regulate the amount of individual pesticides in our
food. However, very little is known about extra-toxic
effects of sub-lethal intake of mixtures of pesticides
or about the form and significance of the chemicals as
consumed by animals (e.g. as plant metabolites). The
question of specific modifications to the original
chemical by different plants and specific suscepti-

Contribution No. 78 Analytical Chemistry Research
Service, and No. 546 Plant Research Institute

Bulletin of Environmental Contamination & Toxicology,
Vol. 1, No. 6, 1966, published by Springer-Verlag New York Inc.

bilities of different animals makes the problem even more complex. Such a situation appears to have arisen in the case of maleic hydrazide (MH), 1,2-dihydro-3,6-pyridazinedione, which is used in agriculture as a plant growth regulator.

While it has been found that MH per se is relatively non-toxic to rats and mice (1), Fischnich et al. (2) reported a significant reduction in the fertility of rats fed with potato tubers from plants sprayed with MH before harvest compared with rats fed a diet with tubers treated with MH during storage. Furthermore, Robinson (3) found a large reduction in the fecundity of pea aphids reared on broad beans grown in soil treated with MH compared with those reared on plants freshly dipped or sprayed with MH. Yule et al. (4), using sexual insects (Musca domestica L. and Drosophila melanogaster Meig.), tested the hypothesis that plant metabolites resulting from treatment with MH could interfere with animal reproduction and found no effects with above-tolerance doses of MH translocated in beans. These authors also confirmed Robinson's findings with MH and pea aphids (3), but favoured the proposal of an indirect nutritional effect through the treated plant as made by van Emden (5).

The present work using insects and potatoes was designed to test if the positive toxicological results with rats and potatoes (2) might be due to plant-specific metabolites of MH.

Experimental

Potato plants (Solanum tuberosum L., cv. Kennebec) were sprayed 5 weeks before harvest with an aqueous solution of MH-30 containing 0.75% a.i. (cf. reference 2). Fresh potatoes were taken from cold storage during the following winter as required to prepare fresh batches of Drosophila rearing medium. Fresh, boiled or freeze-dried tubers were substituted for the whole dry weight of corn meal in a standard diet (6). Ten female and 10 male 1-day-old flies were transferred from a breeding stock of D. melanogaster into a 12-ounce glass jar containing approximately 100 g. of media, and reared and counted as described previously (4).

Three replicated treatments were included in three experimental series: standard medium containing 50 g. fresh, boiled potatoes per 100 g. medium (CP);

CP with 1 to 4 ml. of an aqueous solution of 0.25% W/V
pure MH freshly added, giving a concentration in the
medium of 25-100 ppm MH (CP + MH); standard medium
with MH-treated potatoes added (MHP). From chemical
analysis (7), the average MH content of fresh tubers
was found to be 57.6 ppm MH, the MH content was not
reduced on boiling, and the wet medium contained
approximately 30 ppm MH. In one series of experiments
10 g. freeze-dried potatoes were added per 100 g. wet
medium giving an MH content of approximately 25 ppm.

Results

In three series of experiments each comprising 3
inbred generations of flies per medium type, no significant reduction in fecundity was found comparing a
range of doses of freshly-added MH and potato-trans-
located MH (Table 1). The sole exception to this was
in series A(P), which was the first experiment conducted and whose large variability may reflect inex-
perience which subsequently decreased with practice.
The addition of MH in either form may have slightly
affected fly fecundity compared with CP, as 4/9 of the
experiments showed a significant difference between CP
and one of the other pairs of treatments. However, no
significant difference between any pair of treatments

TABLE 1

Total Numbers of Adults Produced from 10 ♀ Breeding in
Different Larval Media

| | Genera-tion | Number of progeny | | | No. of reps. | MH added[*] (mg.) |
		CP	CP + MH[*]	MHP		
	P	870[a,b]	898[b]	475[a,b]	5	3.5
A	F_1	917[b]	601[b]	816	5	10.0
	F_2	486	525	373	2	3.5
	P	674[b]	824	897[b]	5	10.0
B	F_1	610[b]	686	958[b]	3	3.5
	F_2	643	674	576	3	2.5
	P	1054[a]	954	982[a]	3	2.5
C	F_1	822	739	833	3	2.5
	F_2	735	766	788	3	2.5

Treatments
CP untreated potatoes
CP + MH untreated potatoes + pure MH/100 g.[*]
MHP Potatoes field-treated with MH

Series
A and B fresh, boiled potatoes
C freeze-dried potatoes

[a] F test, variance ratio significant at $P = 0.05$
[b] t test, difference of means significant at $P = 0.05$

was found in the most likely tests (F_2).

Discussion

Drosophila bred well in the potato-agar medium,
development time was normal and moulds presented no

difficulty.

Our conclusion from insect tests is that translocated MH in bean plants (4) and in potato tubers has no deleterious effects on animal reproduction. The question of a specific susceptibility of rats towards translocated MH remains to be resolved.

Acknowledgments

We acknowledge with appreciation the technical assistance of Mrs. D. Turton and thank the Food Research Institute of the Canada Department of Agriculture for supplying potatoes.

References

1. J. M. BARNES, P. N. MAGEE, E. BOYLAND, A. HADDOW, R. D. PASSEY, W. S. BULLOUGH, C. N. CRUICKSHANK, M. H. SALAMAN, and R. T. WILLIAMS, Nature 180, 62 (1957).

2. O. FISCHNICH, C. PAETZOLD, and C. SCHILLER, European Potato J. 1, 25 (1958).

3. A. G. ROBINSON, Can. Entomol. 92, 494 (1960).

4. W. N. YULE, E. V. PARUPS, and I. HOFFMAN, J. Agr. Food Chem. 14, 407 (1966).

5. H. T. VAN EMDEN, Nature 201, 946 (1964).

6. W. N. YULE, Can. Entomol. 97, 269 (1965).

7. I. HOFFMAN and R.B. CARSON, J. Assoc. Offic. Agr. Chemists 45, 788 (1962).

A Method for the Determination of Pentachlorophenol in Human Urine in Picogram Quantities

by A. Bevenue, J. R. Wilson, E. F. Potter and Moon Ki Song
Pacific Biomedical Research Center
University of Hawaii, Honolulu, Hawaii

and

H. Beckman and G. Mallett
Agricultural Toxicology and Residue Research Laboratory
University of California, Davis, California

Pentachlorophenol (PCP), together with its salts, is used as a contact herbicide and defoliant, as a control against micro-biological attack in the manufacture of cellulosic products, adhesives, and paints, as a fermentation inhibitor for non-edible materials, and as a control against termites, powder post beetles, and other wood-boring insects (1, 2, 3). This pesticide can be harmful to humans, especially to the occupationally exposed worker, and acute toxicity problems have occurred with the use of PCP formulations primarily through carelessness or failure to observe the manufacturers' precautionary instructions for its use (4, 5, 6). No chronic toxicity data of any consequence exist on the human individual that may have been exposed to PCP over a long period of time, either through occupational use or through occasional use in the home to control insect infestation. Any

---

The research upon which this publication is based was performed pursuant to Contract No. PH 86-65-79 with the U.S. Public Health Service, Department of Health, Education, and Welfare. The views expressed herein are those of the investigators and do not necessarily reflect official viewpoint of the Public Health Service.

257

Bulletin of Environmental Contamination & Toxicology,
Vol. 1, No. 6, 1966, published by Springer-Verlag New York Inc.

effort to establish a relationship between long-term exposure of the individual to this pesticide and any apparent adverse physiological effects will necessitate the acquisition of analytical data on PCP residues in the human system. Unfortunately, the present available methods of analysis for PCP residues in tissues, and related materials, are applicable only if PCP is present in relatively large amounts (microgram or milligram quantities), or if a large sample is available for analysis (7, 8, 9, 10). The sample size of the material and the detection limit of the analytical procedure will govern the positive or negative findings for any pesticide residue. For example, Gordon (4) suspected PCP poisoning in an autopsy case but failed to find any evidence of PCP residue in the tissue samples with the analytical procedures available at that time. In a subsequent autopsy case, in which PCP poisoning was also suspected, he used much larger quantities of tissues for analysis and obtained positive evidence for the presence of this pesticide in the tissues.

Tissue material cannot be readily obtained from the live individual, and blood and urine specimens are considered to be reasonable substitutes for many clinical and chemical analytical studies. However, it is impractical and often impossible to obtain sufficient amounts of blood and/or urine to utilize the aforementioned analytical methods for PCP. It has been suggested that 24-hour urine samples should be used for analysis. This would be applicable to hospitalized cases, but it would not be feasible for a cross section sampling of a given population. Therefore, a procedure was devised whereby twenty-five to fifty ml of urine was sufficient for the determination of any PCP residues in the picogram-nanogram range.

Experimental

Apparatus

Gas Chromatographs: MicroTek MT 220, electron capture detector, inlet temperature 180 C, column 145 C, detector 193 C; nitrogen flow rate 70 ml/min.

F&M Model 720, TC detector (190 ma), inlet temperature 250 C, column 175 C, detector 270 C; nitrogen flow rate 40 ml/min.

Dohrmann Microcoulometer C-200, Chloride Cell T-300.

Gas Chromatograph Columns: MicroTek and Dohrmann: 5% QF-1 on Gas Chrom Q (100/120 mesh); 10% DC-200 on Chromport XXX (60/80 mesh); 6' x 1/4" glass. F&M: 20% Dow-11 on Chromosorb P (60/80 mesh); 2' x 1/4" stainless steel.

Reagents: Sulfuric acid, redistilled petroleum ether (b. p. 30-60 C).

Diazomethane in ether was prepared from "Diazald" according to directions of the manufacturer (Aldrich Chemical Co., Inc., Milwaukee, Wisconsin), but modified by using 7 g instead of 21.5 g of Diazald and making the final volume of the diazomethane solution to 400 ml with ethyl ether.

Procedure: Twenty five ml urine was transferred to a 125-ml

glass-stoppered erlenmeyer flask. Five ml concentrated sulfuric
acid was added slowly to the flask, with constant mixing of the
contents to avoid superheating the mixture. (If urine samples
are stored under refrigeration prior to analysis, a precipitate
may appear in the samples. Before using, concentrated sulfuric
acid should be added to the entire amount of urine sample in the
same ratio as described above, 5:25, to redissolve the precipitate.) Fifteen ml petroleum ether was added to the erlenmeyer
flask and shaken for two minutes, using a Burrel Wrist Action
shaker. The mixture was transferred to a 125-ml separatory funnel, two ml isopropyl alcohol was added to prevent the formation
of an emulsion, and the lower aqueous phase was removed and transferred back to the original erlenmeyer flask. A second portion
of 15 ml petroleum ether was added to the flask and the shaking
and extraction procedures were repeated. The two petroleum ether
fractions were combined and washed twice with distilled water.
The ether was removed by evaporation with the aid of a stream
of nitrogen at room temperature or, preferably, by allowing the
solutions to stand overnight in beakers, at room temperature, in
a vented area. Accelerated drying on a steam bath with a stream
of air will create a water condensation problem with the residue,
and may necessitate further manipulation steps to remove the

Representative products and manufacturers are named for identifi-
cation only and listing does not imply endorsement by the Public
Health Service and the U.S. Department of Health, Education, and
Welfare.

water, thereby increasing the chances of losses of PCP residues by volatilization. A constant volume of diazomethane solution (5 ml) was added to each sample residue. After a period of 5 minutes, the solvent was removed with a gentle stream of nitrogen and a minimum of heat. The residue was dissolved in petroleum ether, transferred to a 25 ml volumetric flask, made to volume, and aliquots were removed from this solution for PCP analysis by gas chromatography.

Increments of a PCP ether solution were applied to a gas chromatograph equipped with an electron capture detector. The linear range, with an input attenuation of 10 and an output attenuation of 64 (MicroTek GC, full scale sensitivity 3.2×10^{-9} amperes), was between a minimum of 30 picograms and a maximum of about 400 picograms. The gas chromatographic records of the PCP residue in the urine samples were excellent. Background interference from other components of the urine was practically nil, because of the extreme sensitivity of the detector to the PCP ether, and because of the method of preparation of the urine extract. The retention time of the PCP ether on the QF-1 column was 1.5 min; on the DC-200 column 6 min. Additional confirmation was obtained with the F&M gas chromatograph, using a Dow-11 column, and also with the Dohrmann microcoulometer gas chromatograph, which is specific for the chloride component of the PCP molecule.

Duplicate standards and, when possible, duplicate urine samples were used to test the validity of the outlined procedure (see

Table I). Analyses were also made on urine samples fortified with PCP (see Table II). The PCP was dissolved in dilute (ca 0.01N) sodium hydroxide solution, and the resultant sodium pentachlorophenate was used for the preparation of the standard solutions and for the fortification of urine samples. The basic solutions were treated with sulfuric acid and prepared for analysis in the same manner as described above.

TABLE I

Reproducibility of Measurements of Pentachlorophenol in Urine

Picograms PCP in 5 ul Urine*	Mean
8.0 - 11.0	9.5
10.5 - 10.5	10.5
12.0 - 12.0	12.0
18.0 - 18.0	18.0
17.5 - 19.0	18.2
20.0 - 27.0	23.5
25.5 - 27.5	26.5
26.0 - 27.0	26.5
27.5 - 30.0	28.7
30.0 - 32.5	31.2
47.0 - 52.0	49.5
65.0 - 65.0	65.0
68.0 - 74.0	71.0
70.0 - 75.0	72.5
74.0 - 75.0	74.5
85.0 - 90.0	87.5
100.0 - 107.0	103.5
120.0 - 125.0	122.5
143.0 - 147.0	145.0
153.0 - 158.0	155.5
175.0 - 187.0	181.0
212.0 - 216.0	214.0

Standard Deviation ± 5.00%

* Duplicate measurements of PCP found in urine in a random sampling of the local male population.

TABLE II

Recovery of Pentachlorophenol from Urine

Amount PCP Found in Urine* (micrograms)	Amount PCP added (micrograms)	Amount PCP Recovered (micrograms)	% Recovery
0.275	0.50	0.625	80.6
0.360	0.50	0.810	94.2
0.540	0.50	0.990	95.2
0.600	0.50	1.000	90.9
0.625	0.50	1.125	100.0
0.900	0.50	1.255	89.6
1.125	0.50	1.440	88.6
1.175	0.50	1.575	94.0

* 25 ml samples   Average   91.6

Mixtures of urine (25 ml) and sulfuric acid (2 ml) were boiled one hour under reflux conditions; PCP recovery was poor, because the phenol volatilized from the boiling mixture and deposited on the walls of the water-cooled condensers. Comparative PCP analyses were made of urine samples, prepared as described in the proposed procedure above, and of samples hydrolyzed one hour with sulfuric acid in a closed system using "cold finger" condensers. The hydrolyzed samples gave slightly higher PCP values, but the difference was not great enough to warrant using this more tedious procedure.

The identity of PCP isolated from the urine was confirmed by paper chromatography (11), using Whatman No. 1 paper impregnated with 10% paraffin oil and silver nitrate reagent, and developed with acetone-water (70:30). Ultraviolet examination of the paper revealed PCP ether spot areas, which were superimposable on

PCP ether standards spots with an Rf value of 0.21. However, because detection sensitivity had a lower limit of 0.2 micrograms, composite samples were necessary for confirmation. Thin layer chromatography, using silica gel and Rhodamine B, and developed in heptane-acetone (98:2), also confirmed the identity of PCP. Final confirmation was obtained by infrared absorption analysis of sample fractions from the gas chromatograph.

Discussion

The urine samples were treated with an excess amount of acid, to insure complete conversion of any pentachlorophenate in the urine to pentachlorophenol, to possibly eliminate any conjugated phenolic complex that might be present, and to avoid emulsion problems during the preparation of the sample extract. PCP is a relatively stable compound and, because of its chemical structure, coupling or substitution reactions common to many phenols should not occur.

The use of the PCP ether was advantageous for gas chromatography (12), because PCP, _per se_, chromatographs very poorly, if at all, unless specialized column techniques are employed (13, 14). PCP may be oxidized to chloranil (tetra-chloro-p-quinone) (15, 1) and misinterpretation of the gas chromatographic data is possible (16). However, the conditions of the designed method should not produce this compound and supplementary procedures confirmed the fact that PCP, _per se_, was isolated. The methyl ester of 2,4-D, if present as a contaminant, would cause misinterpretation (17), if only the QF-1 column were used; supplementary use of the DC-200 column would eliminate the doubt. The possibility of confusing the PCP ether with the methyl ester of DDA (bis[p-chlorophenyl]acetic acid) was eliminated because of the operating conditions of the gas chromatograph, i.e., the relatively low column temperature increased considerably the retention time of DDA as compared to the early emergence of the PCP ether from the column.

The infrared spectra of a standard PCP ether sample and the material extracted from the urine were prepared in an identical manner. A comparison of the spectra showed that the material extracted from the urine contained all of the major absorption bands observed in the standard spectra. The extracted material also showed absorption bands at 1500 cm^{-1}, 1800 cm^{-1}, 2800 cm^{-1}, and 3000 cm^{-1}. It is the opinion of the authors that these additional bands were derived from aliphatic contaminants.

No significance can be attached, at this time, to the amounts of PCP residue found in the urine of the individuals reported in Table I. This may be possible only after additional residue data have been obtained, together with complete and precise exposure histories of the individuals. It is not known whether the amounts of PCP residues thus far observed in the human system are capable of producing long-term harmful effects. Extensive clinical and toxicological data will be required before these factors can be given consideration. To date, the emphasis of the problem was on analytical methodology.

References

1. MONSANTO CHEMICAL CO., St. Louis, Mo., Tech. Bull. No. SC-8, March 1963.
2. MONSANTO CHEMICAL CO., St. Louis, Mo., Tech. Bull. No. SC-3, October 1958.
3. T. S. CARSWELL and H. K. NASON, Ind. Eng. Chem. 30, 622-626 (1938).
4. D. GORDON, Medical J. Australia, 485-488 (September 29, 1956).
5. DOW CHEMICAL CO., Midland, Michigan, Dowicide Products Bulletin, 1962.
6. MONSANTO CHEMICAL CO., St. Louis, Mo., Tech. Bull. No. SC-9, January 1960.
7. T. AKISADA, Bunseki Kogaku, 14, 101-105 (1965).
8. T. TSUDA and T. KARIYA, Bull. Jap. Soc. Sci. Fisheries 29, 828-833 (1963).
9. K. ERNE, Acta pharmacol. et toxicol. 14, 158-172 (1958).
10. W. DEICHMANN and L. J. SCHAFER, Ind. and Eng. Chem., Anal. Ed. 14, 310-312 (1942).
11. L. C. MITCHELL, J. Assoc. Offic. Agr. Chemists 40, 294-302 (1957).
12. J. KANAZAWA, Agr. Biol. Chem. (Japan) 27, 153-158 (1963).
13. J. R. SMITH, R. O. C. NORMAN, and G. K. RADDA, J. Gas Chromatography 2, 146 (1964).
14. R. H. KOLLOFF, L. J. BREUKLANDER, and L. B. BARKLEY, Anal. Chem. 35, 1651-1654 (1963).
15. M. KUMAHARA, N. KATO, and K. MUNAKATA, Agr. Biol. Chem. (Japan) 29, 880-882 (1963).
16. S. KAWAI, T. KONDO, and T. TOKIED, Eisei Shikenjo Kenkyu Hokoku 81, 49-50 (1963).
17. J. KANAZAWA, Japan Analyst 14, 481-483 (1965).

INDEX to Volume 1